North American Mixing Forum

Presentations at the 2009 AIChE Annual Meeting

Nashville, Tennessee, USA
8-13 November 2009

ISBN: 978-1-61567-930-0

Printed from e-media with permission by:

Curran Associates, Inc.
57 Morehouse Lane
Red Hook, NY 12571

CURRAN ASSOCIATES INC.
proceedings
.com

Some format issues inherent in the e-media version may also appear in this print version.

Printed by Curran Associates, Inc. (2010)

AIChE
3 Park Avenue
New York, NY 10016-5991

Phone: (203) 702-7660
Fax: (203) 775-5177

www.aiche.org

Additional copies of this publication are available from:

Curran Associates, Inc.
57 Morehouse Lane
Red Hook, NY 12571 USA
Phone: 845-758-0400
Fax: 845-758-2634
Email: curran@proceedings.com
Web: www.proceedings.com

TABLE OF CONTENTS

Single and Multiphase Mixing in Stirred Vessels with Low Height-to-Tank Diameter Ratio Using Computational Fluid Dynamics and Particle Image Velocimetry 1
Shilan Motamedvaziri, Piero M. Armenante

Flow Structure within a Jet-Stirred Cylindrical Vessel 2
Khaled J. Hammad, Ivana M. Milanovic

Mixing Prop Optimized for Collimated Flow 12
Bob Higbee, Binxin Wu, Wojtek Wyczalkowski

In-Line Blending of Liquids with Different Densities and/or Viscosities 13
David A. R. Brown, Michael K. Dawson

Effect of a High-shear Homogenizer on Blend Time in a Stirred Tank 14
Micaela Caramellino, Piero M. Armenante

Determination of Pumping Number, Momentum Number and Velocity Profiles of PBT and KPC Impellers 15
Marcio B. Machado, Celso F. Joaquim Jr., Jose Roberto Nunhez, David Nobes, Suzanne Kresta

Gassed Power, Hold-up and Mass Transfer Under High Intensity Sparging and Agitation; Issues with Heterogeneous Gas Flows 16
Alvin W. Nienow, Waldek Bujalski, Klaus Gezork, Michael Cooke

Single Pass Drop Size Distributions Produced by a Multistage in-Line Rotor-Stator Mixer 17
B. N. Murthy, Kenneth T. Kiger, Richard V. Calabrese

A Model for Gas-Liquid Contact in Stirred Vessels 18
Subbarao Duvvuri, Ye Lwin, Dhanshri Sable

An Investigation Into "Just Draw-Down" Conditions for Immiscible Liquid Systems 29
David A. R. Brown

Prediction of Phase Inversion in Oil-Water Systems 30
Krzysztof Piela, Gijs Ooms, Jan V. Sengers

Fabrication of Monodisperse Microcapsules with Stimuli-Responsive Membranes in Microfluidic Devices 33
Liang-Yin Chu

Design Mixers to Minimize Erosion and Corrosion-Erosion 34
Julian B. Fasano, Kevin J. Myers, Eric E. Janz

Successes and Emerging Issues in Simulating the Mixing Behavior of Liquid-Particle Nuclear Waste Slurries at the Savannah River Site 48
David C. Koopman, Bradley R. Pickenheim, Daniel P. Lambert, Michael E. Stone

Solids Suspension by Hydrofoil Impellers Operating Close to the Vessel Bottom at Low Liquid Levels 55
Bob Dowd, Chris Hibshman, Wojtek Wyczalkowski

Solids Suspension of a Mixture of Solids with Two Different Densities in Mechanically Agitated Tanks – Part 1- Low Concentration Effects 56
Arthur W. Etchells III, Robert Hesketh, Maria Garcia-Barberena, Muhammad Usman Iftikhar, Cameron A MacGregor, Keith Weigand

Importance of Particle Properties on the Dispersion of Nanoparticle Clusters 57
Gustavo A. Padron, Gul Nerime Ozcan-Taskin

Particle Transport in Closed Streamline Flows at Small Reynolds Number 58
Jeffrey F. Morris

Temporally and Spatially Resolved Calculations of Particle Mixing in a Coal Jet Using Large Eddy Simulations (LES) and Direct Quadrature Method of Moments (DQMOM) 59
Philip J. Smith, Charles Reid, Jeremy N. Thornock

Mixing and Shear-Induced Migration in a 2D Time-Periodic Cavity-Driven Flow 68
Bu Xu, James F. Gilchrist

Particle Residence Time Distribution in a Continuous Flow Stirred Reactor: Experimental Characterization and CFD Modeling 69
Madan Somasi, Avinash Khopkar, Kishore Kar, Nilesh Parmar, Ed Madenjian

Numerical Simulation of Marangoni Convection with Lattice Boltzmann Method Based On the Field Mediators 70
Yumei Yong, Chao Yang, Yi Jiang

CFD Modelling for Gas-Liquid Two Phases of Single and Dual Impellers System 72
Ruey Chi Hsu, Chung Tung Wu

Aerobic Fermenter Scale-up: Why Equal Mass Transfer Scale-up Doesn't Always Work 73
Julian B. Fasano, Eric E. Janz, Kevin J. Myers

Mixing Time in a Stirred, Torispherical-Bottomed Tank Equipped with a Retreat-Blade Impeller Under Unbaffled and Partially Baffled Conditions 85
Nonjaros Chomcharn, Piero M. Armenante, Billy R. Allen

Computational Fluid Dynamics Simulation of a Quadrupole Magnetic Sorter Flow Channel 86
V. S. K. Sajja, David J. Kennedy, Paul W. Todd, Thomas R. Hanley

Development of a Mixing Model in the Scale-up of An API Crystallization 87
Adam M. Sokolnicki

Micromixing in Stirred Tanks Equipped with a High-Shear Homogenizer 88
Micaela Caramellino, Piero M. Armenante

Influence of Mixing on the Dissolution Performance of a Pharmaceutical Dosage Form 89
Otute Akiti, Sean P. Cusack, Deana Tolentino, Rhea Donley

Mixing Scale-Down Studies for the Scale up of Beer Fermentation 90
Alvin W. Nienow, Georgina McLeod, Christopher J. Hewitt

Comparison of Traditional Rushton Disc Turbines with up-Pumping Hydrofoil B2 Impellers in 550 L Pilot Scale Aerobic Submerged Fermentations 91
Mads O. Albaek, Stuart Stocks, Alvin W. Nienow

Performance Characterization and Scalability of Disposable Bioreactors 92
Ashraf Amanullah, Terrence Allotta, Kapil Kothary, Gayle E. Derfus, Thomas Myint, Terry Hudson

The Simplest Stirred Tank Bioreactor 93
Mario M. Alvarez, David Bulnes-Abundis, Marissa Granados, Josefina Castillo-Reyna

Agitation of Microbial Cultures Via Low-Frequency Acoustic Energy for Improved Oxygen Transfer 102
Todd A. McAdams

Inline Mixing Device for Medical Applications 103
Yves Delmotte

Culture of Human Cells in Asymmetric Laminar Flow Systems 138
Mario M. Alvarez, Leydi Maribel Carrillo-Cocom, David Bulnes-Abundis, Gayathree Murugappan, Keith Johnson

Comparison of Computed and Measured Wall Shear Stress in an Orbiting Cell Culture Dish 145
Jonathan Thomas, M. Keith Sharp, R. Eric Berson

Heat Exchange in the Rotated Arc Mixer: Experiment, Theory, and Plant Trials 146
Guy Metcalfe, Daniel Lester, Murray Rudman

Energy Savings and Improved Mixing Performance with Modified Impeller Technology in Paper Pulp Mixing 147
Marc Moseley, Wojtek Wyczalkowski

Using Computational Fluid Dynamics with Experimental Validation to Predict Liquid-Solid Mixing in Vessels Equipped with Pulsed-Jet Mixers 148
Brigette Rosendall, Milorad Dordevich, YoChan Kim, Chris Kennedy

Mixing Performance of a Novel, Continuous Confined Impinging Jets Reactor Using Competitive Reactions 149
Han Zheng, Giuseppe Di Benedetto, Piero Armenante

Direct Flow Simulations of Thixotropic Liquids in Agitated Tanks 150
Jos Derksen

Simulation of Flow Field and Circulation Time Phenomena in a Batch Rotor-Stator Mixer 151
Meng Yang, Richard V. Calabrese

Simplified Model-Based Design and Shape Optimization for T-Shaped Microreactors with Secondary Flow 152
Lin Wang, Osamu Tonomura, Manabu Kano, Shinji Hasebe

DNS of a Turbulent Stirred Vessel with Baffles and a Rushton Turbine 153
Jurriaan J.J. Gillissen, Harry E.A. Van den Akker

Correlation for Yield of Competitive Reactions with Turbulent Mixing 154
Gary K. Patterson

Simplification and Application of An Interfacial Area Concentration Balance by Order Arguments 155
Richard L. Long

Using Powder Surface Energies to Predict Mixing Properties of Model Excipient-API Blends 156
Daniel J. Burnett, Ron Gray, Nathaniel Catron

The Effect of Mixer Properties and Fill Level on Granular Flow in a Bladed Mixer 157
Brenda Remy, Johannes G. Khinast, Benjamin Glasser

Novel Technique for Improving Granular Mixing in Rotating Cylinders 158
Suman K. Hajra, D. V. Khakhar

Mixing Patterns Obtained in an Oscillatory Shear Mixer 159
Luis Guillermo Obregon Quiñones, Carlos Velazquez

Effect of Particle Size Distribution and Weight of Powders in Mixing of Granular Materials Under Periodic Shear ... 160
Luis Guillermo Obregon Quiñones, Carlos Velazquez
Water Quality Analysis Related to Finished Water Mixing in Water Storage Tanks ... 170
Jason A. Heberling
Impeller Design for Anoxic Mixing Services ... 171
Wojtek Wyczalkowski, Bob Higbee, Binxin Wu
Water and Wastewater Mixing ... 172
Michael K. Dawson, David A. R. Brown
Low-Order Modeling for the Effective Cleaning Radius of a Pulse Jet Mixer ... 173
L. J. Peltier, YoChan Kim, Henry Fitzpatrick, Kelly Knight, Brigette Rosendall, Chris Kennedy, Jon Berkoe
Novel Design for a Magnesium-Enhanced Lime Ex-Situ Gypsum Oxidizer ... 174
Chris Hibshman, Bob Dowd, Wojtek Wyczalkowski, Kevin Smith, Carl Larid, Julia Darling
Improved Mass Transfer Efficiency in Wet Limestone Flue Gas Desulfurization ... 175
Chris Hibshman, Bob Dowd, Wojtek Wyczalkowski
Author Index

Single and Multiphase Mixing in Stirred Vessels with Low Height-to-Tank Diameter Ratio Using Computational Fluid Dynamics and Particle Image Velocimetry

Shilan Motamedvaziri, Otto H. York Department of Chemical, Biological and Pharmaceutical Engineering, New Jersey Institute of Technology, Newark, NJ

Piero M. Armenante, Otto H. York Department of Chemical, Biological and Pharmaceutical Engineering, New Jersey Institute of Technology, Newark, NJ

In many industrial applications mixing vessels have liquid height-to-tank diameter ratio H/T equal to, or larger than, 1. However, there are many instances where this ratio is lower than 1, as in all those cases in which the vessel is emptied, Even when H/T<1, sufficient agitation must still be provided in order to attain process objectives. In such cases, the fluid dynamics of even a single-phase stirred liquid can become quite complex, with different regimes possibly existing depending on the geometric characteristics of the system (such as impeller clearance, liquid height, or liquid head above the impeller). The objective of this work is to obtain a minimum liquid level, for standard impeller off-bottom distance (C/T=1/3), where adequate mixing process can still be achieved both in a single liquid phase and solid-liquid suspension. Computational Fluid Dynamic (CFD) simulation as well as Particle Image Velocimetry (PIV) was used here to study the velocity profile in the flat-bottom baffled vessels equipped with a single Ruston Turbine (RT) in different height-to-tank diameter ratios when liquid level is lover than tank diameter. A commercial pre-CFD mesh generator (Gambit 2.4.6) coupled with a CFD software package (Fluent 6.3.26) with a standard k-epsilon model was used. Results were obtained using a multiple references of frames (MRF) approach. Then transient simulation based on sliding mesh technique has been conducted. Experimental work was also conducted in solid suspension containing glass beads. All the experimental and computational studies were conducted with a single liquid phase as well as in solid-liquid suspensions with 0.5% dispersed solid. A comparison of the experimental and computation work is presented.

See more of this Session: Mixing in Single Phase Systems
See more of this Group/Topical: North American Mixing Forum

2009 AIChE Annual Meeting, Nashville, TN, November 8-13, 2009

FLOW STRUCTURE WITHIN A JET-STIRRED CYLINDRICAL VESSEL

Dr. Khaled J. Hammad, Dantec Dynamics, Ramsey, NJ, USA
Dr. Ivana M. Milanovic, University of Hartford, West Hartford, CT, USA

Abstract

Particle Image Velocimetry (PIV) was used to investigate the turbulent flow structure inside a jet-stirred cylindrical vessel. The turbulent submerged jet issued vertically downward from a long pipe that ensured fully developed turbulent flow conditions at the outlet. The Reynolds number based on jet mean exit velocity was 15,000. The effect of nozzle placement within the vessel on the resulting flow patterns was also studied. The measured turbulent velocity fields are presented using Reynolds decomposition into mean and fluctuating components, which, for the selected flow configuration, inflow and boundary conditions, allow for straightforward assessment of turbulence models and numerical schemes. The flow field was subdivided into three regions: the jet, the jet-wall interaction and bulk of vessel. Proper Orthogonal Decomposition (POD) analysis was applied to identify the most energetic coherent structures of the turbulent flow field in the three regions. The frame independent swirling strength vortex identification technique was used to detect the existence and strength of vortical structures in the jet region.

Introduction

Mixing has a variety of technically important applications and was studied in connection with homogenization of physical properties and composition, control of heat and mass transfer and chemical reactions. Depending on the application, mixing may be achieved using jets, impellers, pipelines and static mixers. Jet mixing has been investigated extensively over last half of a century with a focus on design of mixing systems, mixing time correlations and, recently, validation of Computational Fluid Dynamics (CFD) investigations [1-11]. The current experimental database contains case-specific parameters, while information on complex mixing flow patterns is incomplete. More detailed flow field measurements are needed for a more thorough understanding of jet mixing hydrodynamics [9]. Furthermore, validation of CFD predictions and turbulence models requires accurate data from well defined boundary, inflow and outflow test conditions.

When a vertically downward oriented jet is used, three distinct flow regions can be identified within the stirred tank: the jet, the jet-wall interaction, and bulk of the tank regions. In the jet region, the surrounding fluid is entrained into the jet resulting in an overall velocity decay, as the jet moves downstream. The jet velocity decay rate strongly depends on the initial jet profile. Proximity of the jet to the wall leads to the streamline's divergence away from the jet centerline, a complete jet deflection, and the formation of a jet-wall interaction region characterized by strong shear and high velocity fluctuation levels. The bulk of the tank may have low-velocity zones or strong circulation patterns, depending on the balance between jet mixing by diffusion and advection, respectively. Equipment design and related mixing processes can be improved through better understanding of the formation and interaction of flow structures in these flow regions.

Coherent structures (CS) contain a significant fraction of the total turbulent energy and influence entrainment and mixing. Better understanding of the distribution and dissipation of turbulent energy is critical to improved mixing at the molecular level, which is desirable in mixing applications with chemical reactions. Proper Orthogonal Decomposition (POD) is a statistical pattern recognition tool that can be used to identify CS and describe turbulent flow fields. POD was first introduced in turbulence research by Lumley [12]. Lumley defined CS embedded into turbulent flow as the structure with the largest mean-square projection onto the velocity field. POD can be used both as an advanced processing method to reveal the main energetic features by eliminating the low energy motions [13] and as a means to construct low-dimensional dynamical models of CS.

In the present study, particle image velocimetry (PIV) and POD are employed to investigate flow patterns and coherent structures of a jet-stirred tank. PIV yields snapshots of the flow field containing two instantaneous velocity components, which facilitates POD analysis of turbulent energy in the chosen plane. The spatial topology and turbulent energy were investigated at a plane parallel to the jet axis for two symmetric jet placements. The selected flow configuration, inflow and boundary conditions allow for assessment of turbulence models and numerical schemes. The experimental setup and test conditions are described next.

Fig. 1 Experimental setup

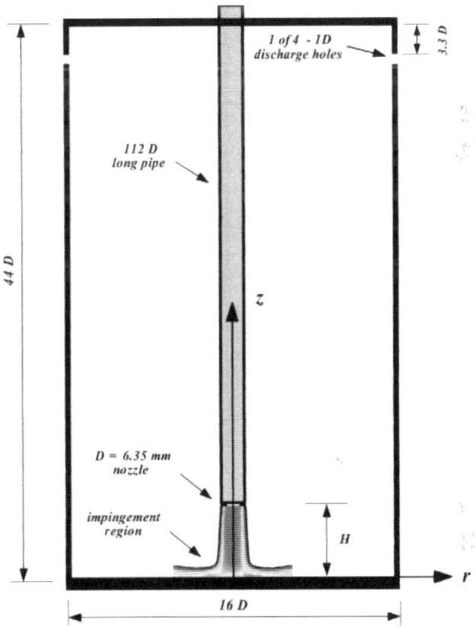

Fig. 2 Test section schematic

Experimental Setup and Test Conditions

Figure 1 illustrates the mixing tank and PIV setup. The closed-loop system comprises of the test section, pump, and the pipe mount. The cylindrical test section, housed in a square tank, is made of smooth finish clear cast acrylic with the diameter $D_{cylinder}$ = 101.6 mm, the height $H_{cylinder}$ = 279.4 mm and wall thickness 3.175 mm. The jet was produced using a submerged vertical precision type 304 stainless steel tube placed on the cylinder centerline with an inner diameter D = 6.35 mm, length L = 711.2 mm and wall thickness 0.381 mm. The length-to-diameter ratio of the tube, L/D = 112, assured a fully developed turbulent flow conditions at the outlet. The resulting $D_{cylinder}/D$ ratio is 16.

As shown in Fig. 2, after impinging on the bottom wall and spreading radially, the overflow leaves the test section through four symmetrically distributed circular holes each with the diameter D_{hole} = 6.33 mm located 20.8 mm below the top plate. The separation distance between the pipe outlet and the tank bottom is denoted with *H*. The origin of the cylindrical *r,z* coordinate system coincides with the center of the tank bottom, or stagnation point. The *z* axis is pointing vertically upward. The *r* coordinate is measured along the bottom circular plate.

Measurements were performed at two separation distances, H/D = 4 and 16. A constant volumetric flow rate of 6.06×10^{-5} m^3/s (0.96 GPM) was used throughout the study, resulting in a fixed pipe bulk velocity, U_b, of 1.91 m/s. The kinematic viscosity of the working fluid, DI water at 30°C, was 8.03×10^{-7} m^2/s. The Reynolds number, $\mathrm{Re} = U_b D/\nu$, was 15,083.

PIV System and Data Processing

A DANTEC DYNAMICS two-dimensional PIV system was used to measure the flow field along each *r-z* plane of interest. It consisted of a dual-cavity Nd:YAG laser with a pulse energy of 120 mJ, 12-bit CCD camera with a resolution of 2048 × 2048 pixels and 7.4 μm in pixel size, synchronization hardware for controlling the laser and camera, and software running on a Windows-based platform for data acquisition, management and post-analysis.

The second harmonic of the Nd:YAG-based laser was used, yielding a pair of pulses of green (532-nm wavelength) beams, fired at a frequency of 7 Hz and typical pulse duration of about 4 ns. Light sheet-forming optics mounted at the exit portal of the laser system generated a thin, focusable light sheet by means of a series of spherical and cylindrical lenses. A vertical divergent light sheet was generated and used to illuminate the full extent of the measurement region within the cylindrical tank. The vertical light sheet was aligned to pass through the center of the pipe. The thickness of the light sheet at the imaging plane was 1 mm. The spatial location of the illumination system was fixed during all the tests.

To eliminate image contamination or saturation due to high intensity reflections near the metal surfaces, fluorescent polystyrene particles were used as the tracer. The particles were doped with Rhodomine B dye, which when illuminated by the green laser light fluoresced in the red, at wavelengths above 560 nm. A high pass color filter (>590 nm) was then used to discriminate particle fluorescence from surface reflections and other Mie-type scattering. Using fluorescent particles to tag the flow resulted in high image contrast and good signal-to-noise performance and allowed for measurements very close to the impingement surface and the pipe tip. The average diameter of the fluorescent particles used was 15 μm, and their density was 1.05 g/cm^3. The latter value was close to the fluid density; hence the particles were nearly neutrally buoyant in the measurement region.

The CCD camera was placed perpendicular to the laser sheet. A micrometer, 0.001 inch positioning accuracy, was used to properly align and optimize the area imaged by the camera. The lens used was a 60-mm *f*2.8 Nikkor. The imaged field of view was 98 × 98 mm, at a magnification coefficient of 0.155.

For the present set of measurements, a high-accuracy multi-pass adaptive cross-correlation technique was used to evaluate the velocity field determined from each pair of particle images. The signal strength was further optimized by window off-set, which minimized particle drop-outs. A sub-pixel accuracy independent of correlation peak shape was achieved and further minimized displacement estimate errors by utilizing an adaptive deforming windows [14]. An initial interrogation area, IA, of 128 × 128 pixels was sequentially reduced to the final IA, 32 × 32 pixels with 50% overlap in both the horizontal and vertical directions. A dynamic second-order accurate method was used to spatially offset the two IA's during each subsequent iteration cycle by the displacement calculated from previous steps.

During each iteration cycle, checks on the correlation's signal-to-noise, as well as a local comparison of vector attributes with the median value of 3 × 3 neighbors were used to validate and remove outliers. To remove spurious vectors, a minimum peak height ratio (between 1st and 2nd peak) of 1.2 was used, thereby putting more stringent conditions on peak identification for the subsequent determination of vectors. 300 PIV realizations were acquired at a sampling rate of 7 Hz for each test case and were used to evaluate the mean and RMS flow field information presented here. Every shown vector represents the spatial average over a 1.5 mm × 1.5 mm area.

Results

The current study focuses on investigating the flow structure in the jet, jet-wall interaction, and bulk of tank regions. Table 1 lists the radial and axial boundaries of these flow regions and number of obtained PIV vectors in both the radial, N_r, and axial, N_z, directions. The pipe exit region was also analyzed to extract the inflow centerline axial velocities, U_c, which were 2.256 and 2.263 m/s for H/D=4 and 16, respectively. U_c = 2.26 m/s was subsequently used to normalize all reported flow field data.

The normalized radial and axial components of the mean velocity are calculated using the following equation:

$$U = \frac{1}{N}\sum_{i=1}^{N}\frac{u_i}{U_c} \tag{1}$$

where u_i is the measured instantaneous velocity and N is the number of PIV data sets for each test case. The fluctuating velocities are obtained after evaluating the mean velocity. Normal and shear Reynolds stresses are calculated using the following equations:

$$\overline{u_r'^2} = \frac{1}{N}\sum_{i=1}^{N}\left(\frac{u_{r,i}}{U_c} - U_r\right)^2 \tag{2}$$

$$\overline{u_z'^2} = \frac{1}{N}\sum_{i=1}^{N}\left(\frac{u_{z,i}}{U_c} - U_z\right)^2 \tag{3}$$

Normalized RMS radial and axial velocity components, $u_{r,rms} = \sqrt{\overline{u_r'^2}}$ and $u_{z,rms} = \sqrt{\overline{u_z'^2}}$, respectively, are calculated and used for data presentation.

Table 1 Summary of boundaries and acquired PIV data for all flow regions

	Jet		Jet-wall		Tank	
	***H/D* = 4**	**16**	**4**	**16**	**4**	**16**
r/D	*-3<r/D<3*	*-3<r/D<3*	*-6<r/D<6*	*-6<r/D<6*	*-6.6<r/D<-1.4*	*-6.6<r/D<-1.4*
z/D	*1<z/D<4*	*1<z/D<15.3*	*0<z/D<2*	*0<z/D<2*	*1<z/D<15.3*	*1<z/D<15.3*
N_r	49	49	99	99	44	44
Nz	23	119	17	17	119	119
vectors	1127	5831	1683	1683	5236	5236

Applying POD to a turbulent flow field consists of finding a series of eigenfunctions or POD modes, with associated turbulent kinetic energy in each mode from a data set. The velocity field can then be expressed as a linear combination of its eigenfunctions. The POD mode represents a CS only if it contains a dominant percentage of energy. POD does not require any prior information on the flow and captures the most energetic components with just a few modes [15]. More information on the POD technique implemented here can be found in Ref. [16]. The energy content of the dominant 10 POD modes was extracted and listed in Table 2. When the jet is located 16 *D* above the tank bottom, POD modes 1-10 contain 30.3%, 46.78% and 62.24% of turbulent energy in the jet, jet-wall, and bulk of tank flow regions, respectively. Moving the jet closer to the tank bottom, *H/D* = 4, results in substantial change in the total energy content of modes 1-10 only for the jet-wall region, 28.11% for *H/D* = 4 vs. 46.73% for *H/D* = 16. This is mainly due to the higher levels of turbulence fluctuations in the jet-wall region obtained when increasing jet-tank bottom separation distance.

Table 2 Energy content of dominant 10 POD modes

	Jet		Jet-wall		Tank	
POD mode	***H/D = 4***	***16***	***4***	***16***	***4***	***16***
1	7.94%	9.43%	5.74%	12.91%	17.94%	27.25%
2	4.48%	5.43%	3.91%	7.82%	14.32%	14.92%
3	4.11%	2.62%	3.35%	6.98%	6.89%	4.98%
4	3.11%	2.24%	3.04%	3.94%	4.59%	3.80%
5	2.85%	2.10%	2.57%	3.02%	3.03%	2.75%
6	2.52%	2.00%	2.42%	2.59%	2.52%	2.58%
7	2.41%	1.81%	2.22%	2.55%	2.21%	1.68%
8	2.26%	1.65%	2.10%	2.21%	1.93%	1.58%
9	1.99%	1.57%	1.92%	1.96%	1.83%	1.42%
10	2.01%	1.44%	1.83%	1.75%	1.58%	1.28%
1-10	**33.68%**	**30.30%**	**29.11%**	**45.73%**	**56.82%**	**62.24%**

1. Jet region

Figure 3 provides a global view of the mean velocity field in the jet region accompanied by the first three POD modes for jet height *H/D* = 16. The mean flow field characteristics display typical submerged jet behaviour, radial growth of the annular shear layer accompanied by decay of axial velocities due to entrainment of the surrounding fluid. The inward penetration of the jet shear layer is complete at about *z/D* = 8. Velocity maps of modes 1-3 contain 9.43%, 5.43%, and 2.62% of the turbulent energy. The higher energy velocity spatial patterns show large scale vortex tubes along the periphery of the jet, from 2-3*D* away from the pipe exit all the way down to the tank bottom and within $-3 < r/D < 3$. Vortex tubes represent various stages of the formation and destruction of jet shear layer vortices.

An overall view of radial and axial turbulence intensities, vorticity and swirling strenght for jet height *H/D* = 16 is shown in Fig. 4. As expected, peak turbulence intensities are observed close to $|r/D| = 0.5$. Radial RMS velocities, $u_{r,rms}$, in the shear layer are higher than those found in the jet core up to about 4*D* away from the pipe. Peak axial RMS values, $u_{z,rms}$, appear in the jet shear layer within 10*D* downstream of the nozzle. The axial turbulence intensity clearly displays the approximately linear growth of the annular shear layer with the downstream distance.

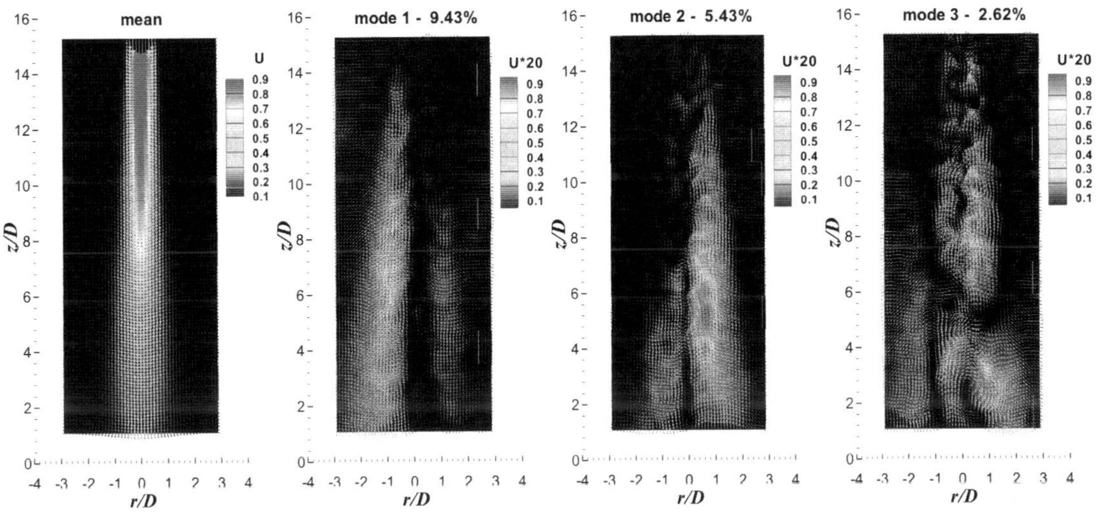

Fig. 3 Flow fields corresponding to mean and dominant POD modes for $H/D = 16$

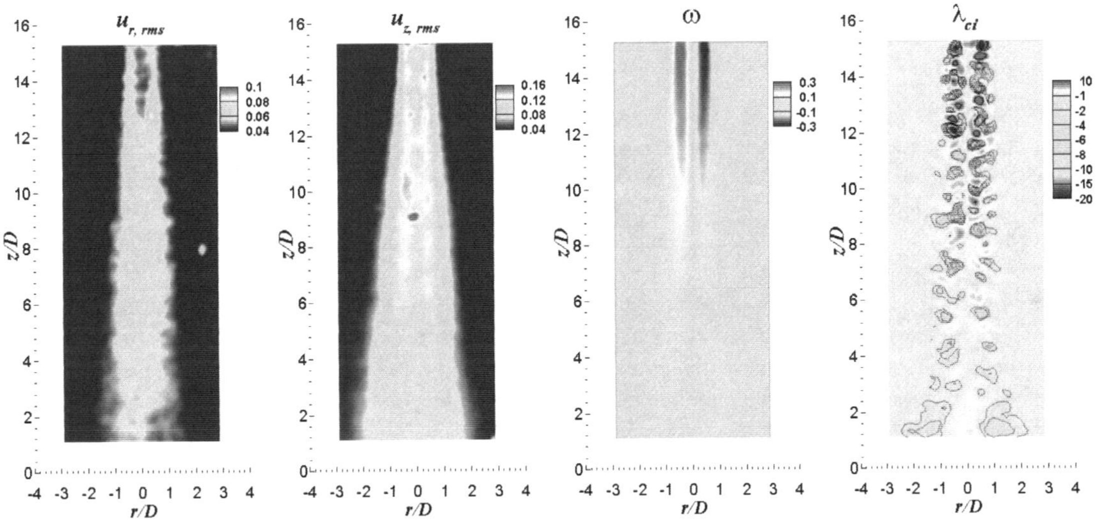

Fig. 4 Turbulence intensity, vorticity, and swirling strength fields for $H/D = 16$

For jet height $H/D = 16$, substantial mean vorticity levels are generated in the shear layer up to $8D$ downstream from the pipe as shown in Fig. 4. Although a number of accepted definitions of vortex make a distinction between a vortex as a structure and a vorticity as a mathematically defined quantity, vorticity analysis has been extensively used to identify location, spread and the 'strength' of vortices. A vortex is typically characterized by high vorticity magnitudes, which is a necessary but not a sufficient condition to detect vortices in both free and wall-bounded shear flows.

Vortices may also be defined as the regions in which the eigenvalues of velocity gradient tensor, $\nabla \boldsymbol{u}$, are complex and the streamline pattern is spiraling or closed in a local reference frame moving with the point. In the case of incompressible fluid, the characteristic equation for the eigenvalues λ of the velocity gradient tensors is given by

$$\lambda^3 + \mathrm{Q}\lambda + \mathrm{R} = 0 \tag{4}$$

where Q and R are the second and third scalar invariants of $\nabla \boldsymbol{u}$. The discriminant, Δ, of the Eq. (4) should be positive:

$$\Delta = \left(\frac{\mathrm{Q}}{3}\right)^3 + \left(\frac{\mathrm{R}}{2}\right)^2 > 0 \tag{5}$$

so that one real eigenpair $(\lambda_r, \boldsymbol{u}_r)$ and a complex conjugate eigenpair $(\lambda_{cr} \pm i\lambda_{ci}, \boldsymbol{u}_{cr} \pm i\boldsymbol{u}_{ci},)$ are allowed. In the regions with $\Delta > 0$, the flow is swirling in the plane of $(\boldsymbol{u}_{cr}, \boldsymbol{u}_{ci})$ around the eigenvector $\boldsymbol{u}_r$. The strength of local swirling motion is quantified by λ_{ci}. Local minima of negative-valued swirling strength identify vortex cores, while positive values indicate areas of the flow where shear may be present. The λ_{ci} fields corresponding to jet height *H/D* = 16 reveal vortical structures throughout the jet region. One diameter downstream of the pipe outlet, an intense and compact vortex pair is observed. Vortical structures in the near field region of the jet engulf the surrounding fluid and substantially contribute to mixing. Further downstream, vortical structures grow in size and diminish in strength, resulting in primarily an inertial and convective mixing.

2. Bulk of tank region

A global view of the mean velocity field in the bulk of tank region accompanied by the first three POD modes for jet height *H/D* = 16 is shown in Fig. 5. Velocity maps of modes 1-3 contain 27.25%, 14.92%, and 4.98% of the turbulent energy. Mean velocity field reveals large scale recirculation pattern with a core at approximately *z/D* = 5 and *r/D* = -5.5. Large scale recirculation is an effective advective mixing mechanism. The large scale vortex is elongated due to the radial spreading of the jet and is located very close to the vertical wall. Substantially reduced velocities are observed in the upper one third and in the bottom left-hand corner of this region which leads to slower mixing. Coherent structures indicate a concentration of turbulent energy in the regions influenced by the jet spread, vortex core, and along the vertical wall slightly above the vortex, resulting in improved mixing due to turbulent diffusion. The bottom left-hand corner is clearly deprived of turbulent energy. This in combination with substantially reduced velocities results in a very slow, diffusion dominated mixing.

The change of the jet height to 4*D* results in somewhat different mean velocity field as seen in Fig. 6. While the large scale vortex is still present at the same radial location, it is also slightly lifted upwards. The jet no longer protrudes, which results in significantly reduced and almost uniform velocities in the lower right-hand region. Stronger recirculation in the lower half and weaker one in the upper one third of the bulk of the tank are also observed. This can be explained by the fact that more of the jet momentum is being transferred to the large scale corner recirculation zone. Velocity maps of modes 1-3 contain 17.94%, 14.32%, and 6.89% of the turbulent energy. The first mode shows energy concentration along the vertical wall only, while the second mode displays energy content trends similar to the 16*D* case. Dominant coherent structures are clearly seen in the left-hand region above the vortex core and around the vortex core.

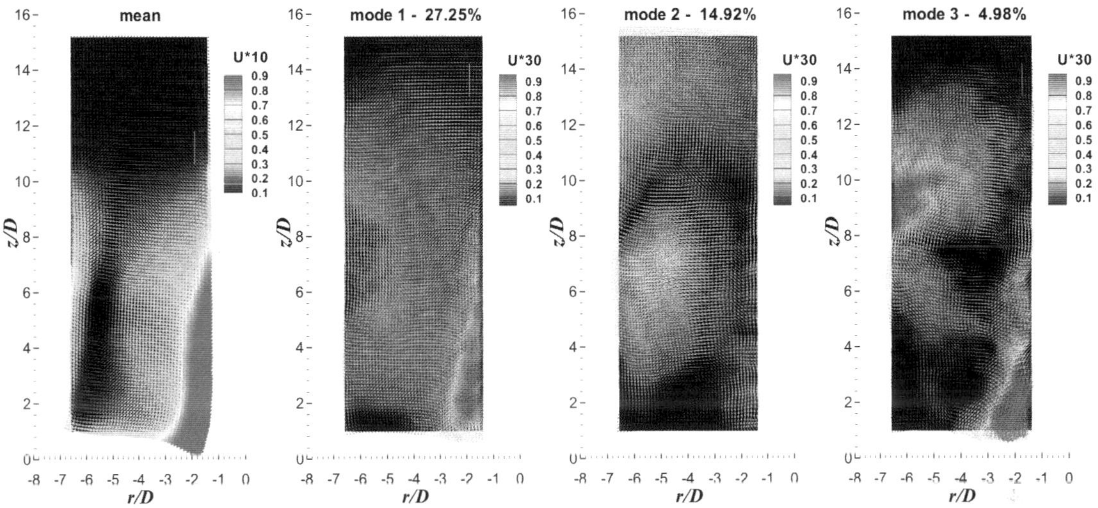

Fig. 5 Flow fields corresponding to mean and dominant POD modes for *H*/*D* = 16

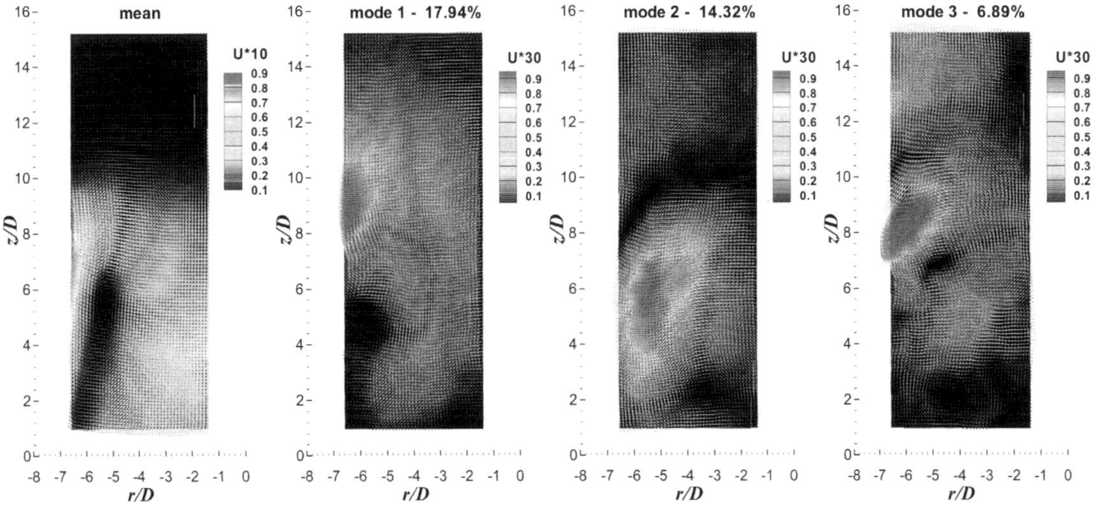

Fig. 6 Flow fields corresponding to mean and dominant POD modes for *H*/*D* = 4

Figure 7 compares radial and axial turbulence intensities for two jet heights, *H*/*D* = 4 and 16. In the case of *H*/*D* = 4, peak turbulence intensities are observed in the recirculation region and above the vortex core, along the vertical wall. However, for *H*/*D* = 16, peak intensities are observed in the bottom right region, due to jet protrusion influence.

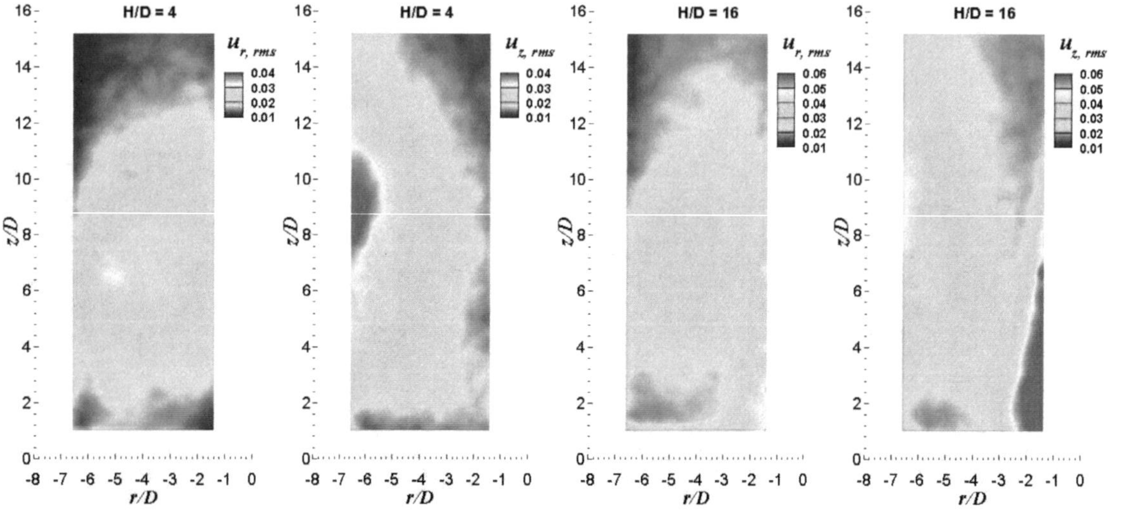

Fig. 7 Radial and axial turbulence intensity fields for *H/D* = 4 and 16

Conclusions

Particle image velocimetry (PIV) and POD were used to investigate flow patterns and coherent structures inside a turbulent jet-stirred tank. Three distinct flow regions were identified and used for presenting the results: the jet, the jet-wall interaction and bulk of tank regions. The spatial topology and turbulent energy were evaluated at planes parallel to the jet axis for two symmetric jet placements, 4 and 16 jet diameters, D, above the tank bottom. The vortical structures throughout the jet region were identified using strength of local swirling motion criterion for the 16*D* case. An intense and compact vortex pair is observed approximately 1D downstream of the pipe outlet. Vortical structures in the near field region of the jet engulf the surrounding fluid and substantially contribute to mixing. Further downstream, vortical structures grow in size and diminish in strength, resulting in primarily an inertial and convective mixing. When the jet is placed 16*D* above the tank bottom, the dominant 10 POD modes were found to contain 30.3%, 46.78% and 62.24% of turbulent energy in the jet, jet-wall, and bulk of tank regions, respectively. Moving the jet closer to the tank bottom, 4D, results in a substantial change in the total energy content of modes 1-10 only in the jet-wall region, 28.11% for *4D* vs. 46.73% for *16D*. Increasing the jet-tank separation leads to higher flow fluctuations close to the tank bottom and coherent structures with higher turbulent energy content. Mean velocity field reveals large scale recirculation pattern in the bulk of tank region. Coherent structures indicate a concentration of turbulent energy around the corner vortex core, the regions influenced by the jet spread, and along the vertical wall slightly above the vortex core. The bottom left-hand corner was deprived of turbulent energy which, in combination with substantially reduced velocities, can result in a very slow, diffusion dominated mixing.

Acknowledgements

The authors are grateful to Mr. Craig Goulbourne for helping with the experiment setup.

References

1. Fosset, H. H., and Prosser, L. E. (1949), 'The Application of Free Jets in the Mixing of Fluids,' *Proc. Of I Mech. Eng.*, 160, pp. 224-251.
2. Fox, E. A., and Gex, V. E. (1956), 'Single-Phase Blending of Liquids,' *AIChE Journal*, 2, pp. 539-544.
3. Hiby, J. W., and Modigell, M. (1978), 'Experiments on Jet Agitation,' *6th CHISA Congress*, Prague.
4. Maruyama, T., Ban, Y., and Mizushina, T. (1982), 'Jet Mixing of Fluids in Tanks,' *J. of Chem. Eng.*, Japan, 15, pp. 342-348.
5. Simon, M., and Fonade, C. (1993), 'Experimental Study of Mixing Performances Using Steady and Unsteady Jets,' *Canadian J. of Chem. Eng.*, 71, pp. 507-513.
6. Orfaniotis, A., Fonade, C., Lalane, M., and Doubrovine, N. (1996), 'Experimental Study of Fluid Mixing in a Cylindrical Reactor,' *Canadian J. of Chem. Eng.*, 74, pp. 203-212.
7. Lane, A. G. C., and Rice, P. (1981), 'An Experimental Investigation of Liquid Jet Mixing Employing a Vertical Submerged Jet,' *I. Chem. Eng. Symposium Series*, 64, pp. K1-K14.
8. Grenville, R. K., and Tilton, J. N. (1996), 'A New Theory Improves the Correlation of Blend Time Data From Turbulent Jet Mixed Vessels,' *Trans. Inst. Of Chem. Eng.*, 74A, pp. 390-396.
9. Jayanti., S. (2001), 'Hydrodynamics of Jet Mixing in Vessels,' *Chem. Eng. Sci.*, 56, pp. 193-210.
10. Patwardhan, A. W., (2002), 'CFD Modeling of Jet Mixed Tanks,' *Chem. Eng. Sci.*, 57, pp. 1307-1318.
11. Zughbi, H., and Rakib, M. A., (2002), 'Investigations of Mixing in a Fluid Jet Agitated Tank,' *Chem. Eng. Comm.*, 189, pp. 1038-1056.
12. Lumley, J. L., 1967, 'The Structure of Inhomogeneous Turbulence,' *Atmospheric Turbulence and Wave Propagation*, ed. Yaglom, A. M. and Tatarski, V. I., Nauka, Moscow, pp. 166-178.
13. Holmes, P., Lumley, J. L, and Berkooz, G., 1996, *'Turbulence, Coherent Structures, Dynamical Systems and Symmetry,'* Cambridge University Press, Cambridge.
14. Westergaard, Madsen, Marassi, and Tomasini, 2003, 'Accuracy of PIV Signals in Theory and Practice,' 5th International Symposium on Particle Image Velocimetry, Busan, Korea, September 22-24, Paper 3301.
15. Gamard, S., George, W. K., Jung, D., and Woodward, S., 2002, 'Application of a Slice Proper Orthogonal Decomposition to the Far Field of an Axisymmetric Turbulent Jet,' Phys. Fluids, **14**, 2515.
16. Hammad, K., and Milanovic, I.M., 2009, 'A POD Study of an Impinging Jet Flow Field,' ASME FEDSM09 Conference, paper FEDSM09-78398, August 2-6, Vail, Colorado.
17. Dalmann, U., 1983, 'Topological Structures of Three-Dimensional Flow Separation,' Deutche Forschungs- und Versuchsanstalt fur Luft- und Raumfahrt Report No. 221-82-A07, Goettingen, Germany.
18. Vollmers, H., Kreplin, H.-P., and Meier, H. U., 1983, 'Separation and Vortical Type Flow Around a Prolate Spheroid - Evaluation of Relevant Parameters,' *Proceedings of the AGARD Symposium on Vortical Type Flow in Three Dimensions*, AGARD-CP-342, Rotterdam, pp. 14-1-14-14.
19. Zhou, J., Adrian, R.J., Balachandar, S., Kendal, T. M., 1999, 'Mechanisms for Generating Coherent Packets of Hairpin Vortices in Channel Flow,' *J. Fluid Mech.*, 387, pp. 353-359.
20. Jeong, J., and Hussain, F., 1995, 'On the Identification of a Vortex,' *J. Fluid Mech.*, 285, pp. 69-94.

Mixing Prop Optimized for Collimated Flow

Bob Higbee, R&D, Philadelphia Mixing Solutions LTD, Palmyra, PA

Binxin Wu, R&D, Philadelphia Mixing Solutions LTD, Palmyra, PA

Wojtek Wyczalkowski, R&D, Philadelphia Mixing Solutions LTD, Palmyra, PA

It has been common practice to borrow marine style props for use in mixing applications even though marine props are not optimized for mixing. Marine props are designed to optimize thrust for a given inlet fluid flow, whereas mixer props must be designed to achieve a desired flow rate and velocity profile. This presentation discusses a propeller design that is optimized for the mixing condition where the prop diameter (D) is small compared to the diameter of the mixing vessel (e.g., 75D). For such applications, the fluid on the inlet side of the prop is virtually stationary (i.e., zero velocity sump). Design goals - Maximize pumping and fluid velocity for a given prop diameter; maximize length of focused, collimated flow; minimize power consumption; and minimize production cost. Prop design employs the novel concept of circular rake along with the established concepts of helical profile, parabolic camber, and skew. A series of (6) Ø8 inch prop CAD models were created using PTC Pro/Engineer software to explore the relationship between three levels of circular rake at two different levels of effective pitch. CAD Models were rendered in glass filled nylon using the selective laser sintering (SLS) rapid prototype process. Performance of CAD models was simulated using Mathcad, Pro/Mechanica (FEA), and Computational Fluid Dynamics analysis. Performance of focused flow impellers was evaluated using Dantec PIV system and simulated in large vessels by CFD. Mixing applications with zero velocity sumps and which require focused flow are discussed.

See more of this Session: Mixing in Single Phase Systems
See more of this Group/Topical: North American Mixing Forum

In-Line Blending of Liquids with Different Densities and/or Viscosities

David A. R. Brown, BHR Group, Cranfield, United Kingdom

Michael K. Dawson, BHR Group, Cranfield, United Kingdom

This paper discusses the performance of a number of static mixer designs in blending miscible liquids which have significantly different densities and/or viscosities. Performance is correlated in terms of required mixing length (number of elements) and mixer densiometric Froude number and modified Reynolds number.

See more of this Session: Mixing in Single Phase Systems
See more of this Group/Topical: North American Mixing Forum

Effect of a High-Shear Homogenizer On Blend Time in a Stirred Tank

Micaela Caramellino, Otto H. York Dept. of Chemical, Biological and Pharmaceutical Engineering, New Jersey Institute of Technology, Newark, NJ

Piero M. Armenante, Otto H. York Department of Chemical, Biological and Pharmaceutical Engineering, New Jersey Institute of Technology, Newark, NJ

Blend time defined as the time required to reach a predetermined degree of homogeneity in a mixing system, is a parameter commonly used to determine the effectiveness of a batch mixing process. Non-dimensional blend times for stirred tanks are available in the literature. However, little information is currently available for fully baffled, stirred tanks equipped with two mixing devices i.e., a centrally mounted slowly rotating impeller to generate bulk flow and a second high-shear homogenizer placed off-center to generate a high-shear zone. These systems are commonly used in the pharmaceutical industry to improve processes such as emulsification and dispersion, which are typically conducted in batch stirred tanks.

In this work, the blend time in a fully baffled, stirred tank equipped with a centrally mounted Rushton turbine as the primary impeller and an additional high-shear rotor-stator homogenizer, was experimentally determined for low-viscosity liquids using a fast acid-base reaction in the presence of an indicator together with an imaging analysis technique to monitor the change in color during the process. Blend time was also determined using conductivity measurements. In addition, the velocity distribution generated by the high-shear homogenizer was experimentally obtained using Particle Image Velocimetry (PIV). These data were used, through a mass balance, to quantify the pumping capacity of the homogenizer, which was compared to that of common impellers. The pumping number for the homogenizer was then calculated at different homogenizer rotational speeds.

Results show a significant reduction in the overall blend time when the high-shear homogenizer is present in the tank, as compared to the blend time obtained with the primary impeller alone. Data were collected at different homogenizer speed and primary impeller speeds, and fitted with a semi-empirical model with only one parameter. The model assumes that the tank is divided into two separated areas that exchange material based on the pumping capacity of the impellers. Data fitting to the model (requiring one fitted parameter to be determined separately and independently) shows a good agreement between the model prediction and the experimental results for blend time under different operating conditions.

See more of this Session: Mixing in Single Phase Systems
See more of this Group/Topical: North American Mixing Forum

Determination of Pumping Number, Momentum Number and Velocity Profiles of PBT and KPC Impellers

Marcio B. Machado, University of Alberta, Edmonton, AB, Canada

Celso F. Joaquim Jr., Engenharia, Kroma Equipamentos Especiais Ltda, Botucatu, Brazil

Jose Roberto Nunhez, UNICAMP, Campinas, Brazil

David Nobes, University of Alberta, Edmonton, AB, Canada

Suzanne Kresta, University of Alberta, Edmonton, AB, Canada

This study aim to examine the performance of an axial impeller, KPC, and a mixed-flow impeller, 45º PBT. The Pumping number, Momentum number and velocity profiles of the impellers were compared. Each impeller was tested with 2 diameters (T/2 and T/3). Two different clearances (C = T/3 and T/4) and two different Reynolds numbers (Re = 30,000 and 40,000) were studied. The data were collected using Particle Image Velocimetry. The circulation patterns were influenced for the impeller shape. The PBT T/2 impeller reached the highest value of axial velocities. The Pumping numbers were influenced for the 3 parameters described in this work: type of impeller (shape and diameter), Reynolds number and clearance. The biggest value of Pumping number (0.71) was reached for PBT T/2 using the following conditions: Re = 40,000 and C = T/3. The smallest value (0.40) was reached for KPC T/3 when the Reynolds number was 40,000; in this case, the pumping number was equal for both clearances. The biggest value of the Momentum number (2.75 N) was reached for PBT T/2 using C = T/3 and Re = 40,000.

See more of this Session: Mixing in Single Phase Systems
See more of this Group/Topical: North American Mixing Forum

Gassed Power, Hold-up and Mass Transfer Under High Intensity Sparging and Agitation; Issues with Heterogeneous Gas Flows

Alvin W. Nienow, Chemical Engineering, University of Birmingham, Birmingham, United Kingdom

Waldek Bujalski, Chemical Engineering, University of Birmingham, Birmingham, United Kingdom

Klaus Gezork, Ekato RMT GmbH, D-79650 Schopfheim, Germany

Michael Cooke, School of Chemical Engineering and Analytical Science, University of Manchester, Manchester, United Kingdom

With high intensity agitation and sparging, heterogeneous flow is found in which much of the gas passes through a reactor in the form of large bubbles (> ~ 30 mm). This condition has been studied with mean specific energy dissipation rates up to 100 W/kg and gas superficial velocities, vs, up to 0.13 m/s (~30 vvm) with a wide range of impeller types. The transition from homogeneous flow occurs at vs between ~ 0.02 and 0.05 m/s and depends mainly on the coalescence characteristics of the fluid and less on scale, the agitator type and mean specific energy dissipation rate. The phenomenon has rarely been studied because it is difficult to do so at pilot scale, though it often occurs at plant scale because with constant vvm (fixed stoichiometry) scale-up and geometric-similarity, vs is proportional to linear scale. Compared to homogeneous conditions, the minimum ratio of gassed to ungassed power is much lower, e.g., Pg/P = ~0.2 with Rushton turbines, ~0.6 with Scaba 6SRGT and ~0.4 with down- or up-pumping high solidity hydrofoils. Hold-up increases much more slowly with vs and may even fall; it is dependent on impeller type, especially with non-coalescing fluids with values up to 0.7. With this hold-up, Pg/Pdisp > 1 for 6SRGT (where Pdisp is based on the dispersion density). Mass transfer coefficients are difficult to measure accurately, especially at low vs and high mean specific energy dissipation rate, because of intense 'surface' gassing. This makes it difficult to determine the relationship between mass transfer and agitation and aeration conditions. Nevertheless, it is suggested that an equation of the usual type can describe the link beteen kLa, vs and mean specific energy dissipation rate where the empirical constants differ from those found for the homogeneous regime, though they are still independent of impeller type and the exponent on vs is less. In addition, the values of these empirical constants are very sensitive to composition, with kLa generally much higher in non-coalescing fluids.

See more of this Session: Mixing in Multi-Phase Systems - Gas-Liquid and Liquid-Liquid Systems
See more of this Group/Topical: North American Mixing Forum

Single Pass Drop Size Distributions Produced by a Multistage in-Line Rotor-Stator Mixer

B. N. Murthy, Chemical & Biomolecular Engineering, University of Maryland, College Park, MD

Kenneth T. Kiger, Mechanical Engineering, University of Maryland, College Park, MD

Richard V. Calabrese, Chemical & Biomolecular Engineering, University of Maryland, College Park, MD

Continuous or in-line high shear mixers are broadly employed in chemical processes to produce emulsions and liquid-liquid dispersions. Despite their widespread use, there is little fundamental basis to theoretically predict or experimentally assess their performance. As a result, process development, scale-up and operation are often by trial and error, leading to higher processing costs, start-up problems and lost time to market. In order to develop a fundamental understanding, at least on a mechanistic basis, of how drop breakup affects liquid-liquid dispersion performance, we have measured power draw and single pass drop size distributions (DSD) exiting a IKA Labor Pilot in-line rotor-stator mixer. This slot and tooth device can accommodate up to 3 stages of generators. We have examined medium (2 rows of teeth per stage), fine (3 rows per stage) and ultra fine (4 rows per stage) generators.

A continuous, turbulent water phase is fed to the mixer via a progressive cavity pump. At time zero, a single large oil drop is injected into the device and the resulting daughter DSD is measured at the exit using our previously reported Phase Doppler Anemometry (PDA) technique. Since PDA yields the time history of the exiting drop population, or residence time distribution (RTD), bivariate statistics of drop size and residence time can be obtained from several parent drop trials. Considerable care has been taken to insure that the PDA yields accurate and reproducible results over a broad range of drop size.

Systematic experiments have been performed to study the effect of flow rate, rotor speed, drop viscosity, generator geometry, and number of stages on the resulting DSD and RTD. Results indicate that flow rate is not as influential as geometry, rotor speed and drop viscosity, especially for multiple stages. The data will be discussed in detail, as well as implications to process performance.

See more of this Session: Mixing in Multi-Phase Systems - Gas-Liquid and Liquid-Liquid Systems
See more of this Group/Topical: North American Mixing Forum

A model for gas-liquid contact in stirred vessels

Duvvuri Subbarao , Ye Lwin, Dhanashri M.Sable
Department of Chemical Engineering,
Universiti Teknologi PETRONAS, Bandar Seri Iskandar, 31750, Tronoh, Perak, Malaysia

Abstract

Gas-liquid contact in stirred tanks received great attention due to their wide industrial applications. Parameters such as power consumption, bubble size, bubble hold-up and volumetric mass transfer coefficients as a function of stirrer speed and gas flow rate were experimentally measured and correlations useful for design were developed. There is growing need to rationalize the vast and wide variety of information available in the literature through application of principles of multiphase flow in stirred tanks. Experimental observations on a gassed stirred tank are presented and analyzed through model based equations. Liquid circulation generated by rotating impeller appears to counter the liquid circulation generated by the flowing gas. The net liquid circulation defines the gas hold-up and power consumption.

Keywords: gassed stirred tank, multiphase hydrodynamics, gas holdup, power draw.

1. Introduction

Gas-liquid reactions such as oxidations, halogenations, hydrogenations, fermentations, polymerizations are carried out at very large scale in stirred tank reactors. Design of such large reactors is still an art and is largely based on not so easily available empirical information and scale-up strategies. Information on parameters of transport phenomena along with reaction kinetics are needed for rational design of such large industrial units. There is a massive body of literature available on most aspects of gas–liquid mixing [1].

Mechanics of single phase fluid motion in stirred relates flow rate to pumping power required to overcome the resistances through the concept of friction factor/power number.

$$N_p = \frac{P}{\rho_l N^3 D^5} = \frac{k_L}{N_{\mathrm{Re}}} + k_T = \frac{k_L}{\left(ND^2 \rho_l / \mu \right)} + k_T \qquad (1)$$

The constants k_L and k_T depend on geometry of the vessel, impeller and other internals. Power draw increases with impeller speed and so it is necessary to identify the minimum impeller speed that can do the required job effectively.

Rate of mixing of fluid is related to liquid circulation time, defined as volume per volumetric flow rate Q_l generated by the impeller, and can be estimated from the liquid flow number

$$N_{Fl} = \frac{Q_l}{ND^3} = k_{Fl} \qquad (2)$$

Identifying the operating conditions and the necessary power input to achieve good gas-liquid contact for mass transfer is the design objective. Some of the typical results in the literature are reviewed here.

a). Flooding:

It has been observed that increase in gas flow rate can lead to a condition known as "flooding" where impeller liquid stream fails to disperse upward rising bubbles in the radial direction (Rushton and Bimbinet [2]). Warmoeskerken and Smith [3] suggested, at the flooding transition, the magnitudes of the radial liquid flow generated by impeller action and axial liquid flows induced by rising bubbles can be expected to be of the same order. Warmoeskerken and Smith [3] obtained a correlation for flooding as

$$\frac{Q_g}{ND^3} = 1.2\frac{N^2D}{g} \qquad (3)$$

Bombac and Zun [4] summarized various correlations available in the literature for estimating flooding condition. Neinow et al. [5] recommended

$$\begin{aligned} \frac{Q_g}{ND^3} &= 30\left(\frac{D}{T}\right)^{3.5}\frac{N^2D}{g} && \text{for flooding - loading transition} \\ \frac{Q_g}{ND^3} &= 0.2\left(\frac{D}{T}\right)^{0.5}\frac{N^2D}{g} && \text{for loading - dispersion transition} \end{aligned} \qquad (4)$$

b) Gas-Liquid Contact in Stirred Tanks

Interfacial area in gas-liquid contactors can be calculated from the values of the average bubble size and gas hold-up, assuming bubble to be spherical, as

$$a = \frac{6\,\varepsilon_g}{D_b} \qquad (5)$$

Gas hold-up, ε_g, can be estimated by using the equation derived by Kudrewizki and Rabe, [6]

$$\frac{\varepsilon_g}{1-\varepsilon_g} = 0.819\frac{u_{og}^{2/3}N^{2/5}T^{4/15}}{g^{1/3}}\left(\frac{\rho_l}{\sigma}\right)^{1/5}\left(\frac{\rho_l}{\rho_l-\rho_g}\right)\left(\frac{\rho_l}{\rho_g}\right)^{1/5} \qquad (6)$$

Bubble size can be estimated from the equation proposed by Calderbank, [7]

$$D_b = 0.7\frac{\sigma^{0.6}}{\left(P_g/V\right)^{0.4}\rho_l^{0.2}} \qquad (7)$$

Mounir Bouaifi and Michel Roustan [8] proposed a correlation for power draw as

$$\frac{P-P_g}{P} = 65\frac{u_{og}}{(gD)^{0.5}} \qquad (8)$$

They also observed that, at any gas velocity, gas holdup correlates well with gassed power per volume.

Multiphase flow through process equipment is extremely complex. Simple equations based on models describing the essential science with in the tractable limits are needed to facilitate design and operation. For example, the correlating approach of Lokhart-Martinelli [9] is still widely used for estimating pressure drop for two phase flow in a pipe from information on pressure drop for each phase flowing in the pipe alone. Wallis [10] showed that it is possible to derive equations to explain the Lokhart Martinelli correlating approach. An attempt is made in this note to explore the possibility of applying the two phase flow approaches suggested by Wallis [10] to stirred gas-liquid contactors.

2. Present Model

From Eq.(1) for power number, power draw for liquid mixing in laminar flow can be expressed as

$$P = k_L \mu N^2 D^3 \tag{9}$$

and for turbulent flow

$$P = k_T \rho_l N^3 D^5 \tag{10}$$

Using this information, an attempt is made to analyze the power draw for gas-liquid mixing in stirred tanks through pseudo-homogeneous flow model and separated flow model.

Pseudo-Homogeneous Flow Model:

A two phase mixture can be considered as a pseudo homogeneous fluid if the two components have the same viscosity, density and velocity; alternatively, to facilitate the assumption of pseudo-homogeneous fluid, the mixture is considered to be well represented by an effective density, viscosity and velocity.

For laminar flow, based on Eq.(9), power draw for the pseudo-homogeneous fluid can be expressed as

$$P_g = k_L \mu_m N^2 D^3 \tag{11}$$

Viscosity of the pseudo-homogeneous fluid is related to dispersed phase holdup in the liquid Wallis [10] . From Eqs.(9) and (11),

$$\frac{P_g}{P} = \frac{\mu_m}{\mu} \tag{12}$$

In a similar way, for turbulent flow

$$P_g = k_T \rho_m N^3 D^5 \tag{13}$$

and the ratio for turbulent flow is

$$\frac{P_g}{P} = \frac{\rho_m}{\rho_l} = (1-\varepsilon_g) \quad (14)$$

Both density and viscosity ratios are related to volume fraction of gas in the stirred tank.

Separated Flow Model

As per this model, liquid and gas bubbles are recognized as two separate phases. Power draw and impeller rotational speed for the mixture is the same for each phase (gas as well as liquid). However, in gas-liquid mixing, the impeller is occupied partly by the liquid phase and partly by the gas phase. Writing equations for liquid phase, for laminar flow, power can be expressed

$$P_g = k_L \mu N^2 D_L^3 \quad (15)$$

From Eqs. (11) and (15), for laminar flow

$$\frac{P_g}{P} = \frac{D_L^3}{D^3} = (1-\varepsilon_g) \quad (16)$$

In a similar way, for turbulent flow

$$P_g = k_T \rho_l N^3 D_L^5 \quad (17)$$

and the ratio for turbulent flow is

$$\frac{P_g}{P} = \frac{D_L^5}{D^5} = (1-\varepsilon_g)^{5/3} \quad (18)$$

Model equations by these two approaches are summarized in Table 1.

Table 1

Flow Model	Laminar Flow	Turbulent Flow
Pseudo-Homogeneous Flow Model	$\frac{P_g}{P} = \frac{\mu_m}{\mu}$	$\frac{P_g}{P} = \frac{\rho_m}{\rho} = (1-\varepsilon_g)$
Separated Flow Model	$\frac{P_g}{P} = \frac{D_L^3}{D^3} = (1-\varepsilon_g)$	$\frac{P_g}{P} = \frac{D_L^5}{D^5} = (1-\varepsilon_g)^{5/3}$

It can be concluded that

$$\frac{P_g}{P} = (1-\varepsilon_g)^n \cong 1 - n\varepsilon_g$$
$$\frac{P-P_g}{P} \cong n\varepsilon_g \quad (19)$$

2.1 Model for gas holdup:

Pseudo-Homogeneous Model:

This model is based on the consideration that the actual velocity of gas and liquid are equal

$$\frac{Q_g}{A\varepsilon_{g.homog}} = \frac{k_1 N D^3}{A(1-\varepsilon_{g.homog})} \tag{20}$$

$$\varepsilon_{g.homog} = \frac{\dfrac{Q_g}{ND^3}}{k_1 + \dfrac{Q_g}{ND^3}} \tag{21}$$

With the constant k_1 assigned a value of 0.4, estimated gas holdup is compared with the measured gas holdup in Fig. (1). It can be seen that the estimated gas holdup by the homogeneous model is many times greater than the experimentally observed gas holdup. This is due to higher velocity of gas phase compared to the velocity of liquid phase due to buoyancy force.

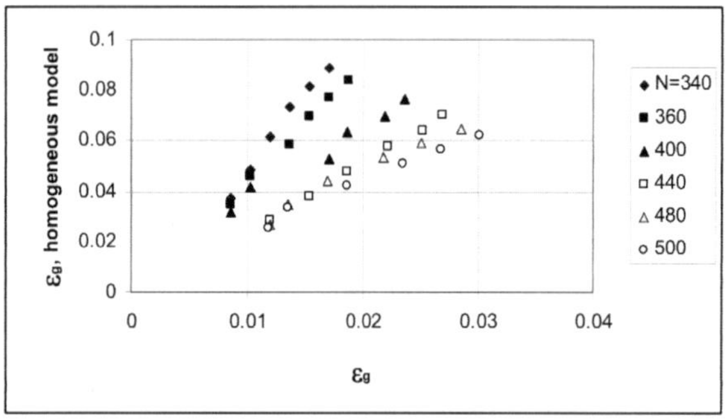

Fig. (1). Comparison of gas holdup estimated from pseudo-homogeneous model with experimental observations.

Drift Flux Model:

From material balance on gas flow through a bubble column, gas holdup can be obtained as

$$u_{og} = u_B\,\varepsilon_g$$

$$\varepsilon_g = \frac{u_{og}}{u_B} \tag{22}$$

A single bubble rises through a pool of liquid due to buoyancy force at a velocity of u_{br}. Generally it is observed that in bubble columns velocity of bubbles is higher than the rise velocity of a single bubble. Nicklin [11] proposed that bubble velocity is enhanced due to incoming superficial gas velocity

$$u_b = u_{og} + u_{br} \tag{23}$$

Based on experimental observations, Zuber and Findlay [12] proposed

$$\varepsilon_g = \frac{u_{og}}{c_0 u_{og} + u_{br}} \tag{24}$$

The constant c_0 is indicative of radial non-uniformity in gas holdup which can generate liquid circulation. Hughmark [13] observed that a value of 2 for the constant c_0 can explain experimental observations in bubble columns.

$$\varepsilon_g = \frac{u_{og}}{2u_{og} + u_{br}} \tag{25}$$

In the present work, it is proposed that observed bubble velocity is due to bubbles rising in a stream of gas and circulating liquid flow

$$u_B = u_{og} + u_{br} + u_{cr} \tag{26}$$

and bubble hold up in a stirred bubble columns can be expressed as

$$\varepsilon_g = \frac{u_{og}}{u_{og} + u_{br} + u_{cr}} \tag{27}$$

2.3 Liquid Circulation velocity:

Bubbles in a pool of liquid rise due to the buoyancy force; the same buoyancy force acts on the liquid as well to set liquid in upward motion along with bubbles in the core region. As the bubbles erupt into the freeboard at the top of the liquid pool, the up flowing liquid in the core circulates down along the wall in the annular region. Rising bubbles tend to move into the core region avoiding the down flowing liquid in the annular region. Considering this phenomena, Freedman and Davidson [14] developed equations to explain the bubble trajectories and strength of liquid circulation in an inviscid liquid. Viswanathan and Subbarao [15,16] extended the model of Freedman and Davidson to obtain a closed form explicit equation to estimate the strength of liquid circulation. Crabtree and Bridgewater [17] and Reitema and Ottengraf [18] developed equations for liquid circulation in bubble columns operating under laminar flow conditions. There has been considerable effort ever since to analyze liquid circulation in bubble columns operating under laminar/turbulent flow conditions with ever increasing complexity to obtain axial and radial velocity profiles. Application of this approach to understand liquid circulation in stirred bubble columns can be formidable due to the complexity of flow created by impellor action. The traditional friction factor concept can be useful to obtain average circulation velocity (Heijnen et al.[19]).

A simple model for liquid circulation velocity in bubble columns:

Liquid in a bubble column is set into circulation by the buoyancy force due to the presence of bubbles acting on the liquid while the frictional drag at the wall resists such circulation. By equating the driving force for flow to the resistance force as

$$\frac{A\left(1-\varepsilon_g\right)}{2} L\varepsilon_g \rho_l g \quad = \quad fA_w \frac{\rho_l u_{crb}^2}{2} \tag{29}$$

Average liquid circulation velocity u_{crb} can be obtained as

$$u_{crb} \quad = \quad \left\{ \frac{Rg\varepsilon_g(1-\varepsilon_g)}{f} \right\}^{1/2}$$

with (30)

$$\frac{1}{R} \quad = \quad \frac{A_w}{AL} \quad = \quad \frac{4}{T} + \frac{1}{L}$$

The factor R is the hydraulic radius. For stirred tanks with liquid depth L equal to T, R is equal to $T/5$ and liquid circulation is controlled by column diameter. The friction factor, f, is due to the wall friction as well as change of direction along the flow path. Assigning a value of 1 to the friction factor f can give reasonable estimates for liquid circulation velocity in bubble columns.

A simple model for liquid circulation velocity in stirred bubble columns

Liquid circulation in stirred bubble columns or gassed stirred tanks is too complex. It is conjectured that liquid circulation created by the impeller action opposes the liquid circulation set in by the bubbles.

$$u_{cr} \quad = \quad u_{crb} - c_1 ND \tag{31}$$

With this gas hold up in gassed stirred tanks can be expressed as

$$\varepsilon_g \quad = \quad \frac{u_{og}}{u_{og} + u_{br} + u_{crb} - c_1 ND} \tag{32}$$

And power draw can be expressed as

$$\frac{P - P_g}{P} \quad = \quad n \frac{u_{og}}{u_{og} + u_{br} + u_{crb} - c_1 ND} \tag{33}$$

3. Experimental Study

The agitated tank used was a flat-bottomed, acrylic glass cylindrical vessel of 28.6 cm i.d. equipped with four equally spaced baffles of width 2.9 cm. A transparent sheet with scale markings was attached to the tank along the height to measure liquid level in the tank. Standard 6 blade Rushton turbine type impellers of three sizes, 14.3, 10.3, and 8.2 cm diameter, were used. Plane of the impeller disk was located at 12.5 cm from the bottom of the tank. Gas was sparged through a simple pipe located 8 cm below the plane of the impeller disc. The working fluids were filtered tap water and compressed air. Liquid level was around 29 cm. Gas flow rate was measured using a rotameter. Impeller rotational speed was measured using an electronic tachometer. The torque is measured by a direct reading force balance dynamometer. The unit was similar to the one used by

Gezorg et al.[20]. Sable [21] reported measurements on power draw and gas holdup as a function of gas flow rate (0 to 0.77 litres/s) and stirrer speed (0 to 500 rpm).

4. Model Evaluation:

Gas Holdup:

The model equation (32) for gas holdup can be rearranged as

$$\frac{1-\varepsilon_g}{\varepsilon_g} = \frac{u_{br} + u_{crb} - c_1 ND}{u_{og}}$$

This equation is tested in Fig.(2) using experimentally observed gas hold up parameter $[(1-\varepsilon_y)/\varepsilon_y]$ as a function of gas superficial velocity parameter $(1/u_{og})$ at various impeller speeds. In the range of gas velocities investigated, the impeller was flooded for stirrer speeds smaller than 300 rpm for impeller of 8.2 cm diameter; the parameter $[(1-\varepsilon_g)/\varepsilon_g]$ is high with a near zero slope. At stirrer speeds greater than 340 rpm, the gas was well dispersed and the slope of the linear relationship between $[(1-\varepsilon_g)/\varepsilon_g]$ and $(1/u_{og})$ (passing through origin as expected from Eq.(32)) is reasonably good. Estimated slopes in the non-flooded region for all the three impellers investigated decreased with tip speed as shown in Fig.(3)

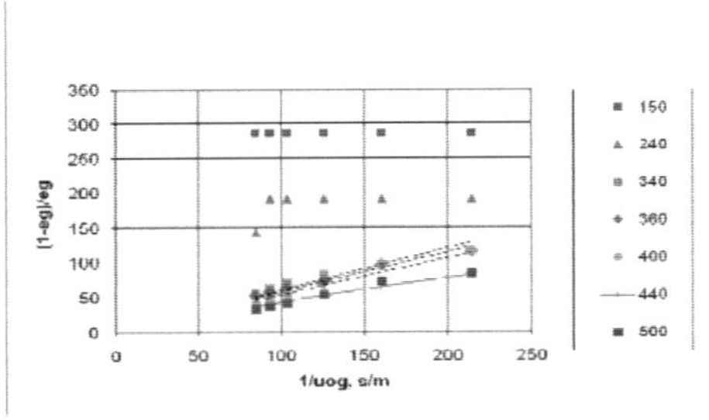

Fig (2). Effect of impeller speed on gas holdup as a function of gas velocity.

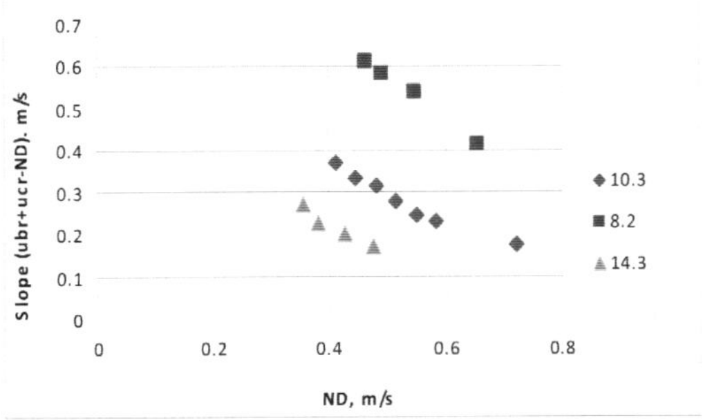

Fig.(3). Estimated bubble velocity as a function of tip speed

The values of the intercept $(u_{cr}+u_b)$ and slope c_1 as a function of impeller diameter are shown in Table 2.

Table 2

D	$u_{cr}+u_b$	c_1
8.2	1.095	-1.0352
10.3	0.612	-0.631
14.3	0.535	-0.7695

The intercept, due to bubble rise velocity and liquid circulation velocity, is smaller for larger diameter impellers. Estimated liquid circulation velocities are in the range of 0.1 to 0.16 m/s. This suggests that bubble rise velocities to be in the range of 0.4 to 0.9 m/s. Experimental observations indicate that most of the bubbles are in the size range of 2 to 3 mm and their rise velocity is expected to be around 0.235 m/s. A few much larger bubbles (> 1cm in diameter) do form by coalescence and rise at much higher velocities; such larger bubbles can entrain smaller bubbles as well (like a swarm) at the velocity of larger bubbles. The smaller intercept for larger diameter impellers indicates smaller rise velocity of large bubbles and hence smaller large bubbles. Larger diameter impellers are observed to produce finer large bubbles.

Power Draw (P_g):

Experimental measurements on $(P-P_g)/P$ are observed to be a function of gas holdup as indicated by Eq.(19); however, there appears to be an influence of scale factor and a reasonable correlation between $(P-P_g)/P$ and $0.4\ (T/D)^3\ e_g$ as shown in Fig.(4). Further detailed work is needed to develop a reliable correlation.

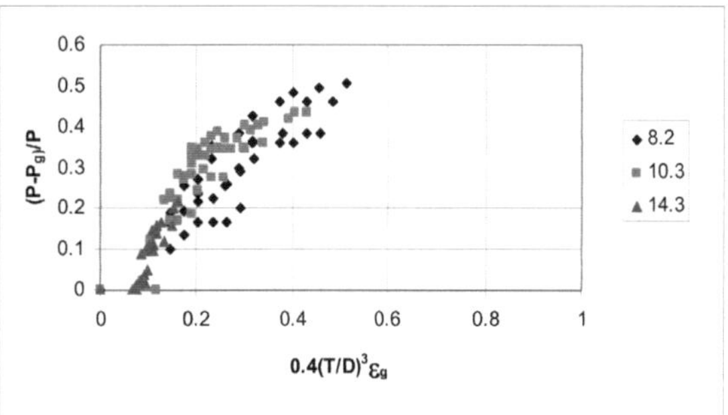

Fig.4. Correlation of Power draw with gas holdup.

5. Conclusions:

a). Based on pseudo-homogeneous model and separated flow models, the factor P_g/P is shown to be a function of gas holdup.

b). Experimentally observed gas holdup is much smaller than gas holdup estimated from the equation developed for gas holdup based on pseudo-homogeneous model.

c). Equation for bubble holdup based on drift flux concept is promising.

d). Bubble swarm velocity may be dictated by the size of large bubbles.

e). Larger diameter impellers appear to reduce diameter of larger bubbles.

Acknowledgement:

The authors thank Universiti Teknologi PETRONAS for the encouragement and financial support for carrying out this work.

Nomenclature:

a	gas-liquid interfacial area
A	Area of cross section of the stirred tank
A_w	Wetted Area of stirred tank
c_o	Constant
D	Impeller diameter
D_b	Diameter of bubble
D_L	Diameter of the impeller occupied by liquid
f	friction factor
g	Acceleration due to gravity
k_1	Constant
k_{fl}	Constant
k_L	Constant
k_T	Constant
L	Liquid height in the tank
N	Stirrer Speed
n	Constant
P_g	Power input under gassed condition
P	Power input with out gassing
Q_g	Gas flow rate
Q_l	Liquid pumping rate of the impeller
R	Hydraulic Radius of the stirred tank.
T	Tank Diameter
u_{br}	Bubble rise velocity
u_B	velocity of bubbles
u_{cr}	net liquid circulation
u_{crb}	liquid circulation in bubble column
V	volume of liquid in the tank
N_{fl}	Flow Number(=Q_l/ND^3)
N_{flg}	Flow Number(=Q_g/ND^3)
N_p	Power Number,(=$P/\rho_l N^3 D^5$)
N_{Re}	Reynolds Number,(=$ND^2\rho_l/\mu$)

Greek Letters

ε_g	Gas holdup
$\varepsilon_{g.homog}$	Gas holdup assuming pseudo-homogeneoud flow
μ	Viscosity of the liquid
ρ_m	Density of mixture
ρ_l	Density of liquid
ρ_g	Density of gas
σ	Surface Tension

References

1. G.B. Tatterson, Fluid Mixing and Gas Dispersion in Agitated Tanks, McGraw-Hill, New York, (1991).
2. Rushton J. H. and Bimbiuet J. J. , Holdup and flooding in air liquid mixing. Can. J. Chem. *Eng. 46* (1968)16-21.
3. Warmoeeskerken. M. M. C. G. and Smith. J. M., Flooding of disc turbines in gas-liquid dispersions: A new description of the phenomenon, Chem. Eng. Sci. Vol. 40. No. I I., (1985), 2063-2071.
4. Andrej Bombac and Iztok Zun, Individual impeller flooding in aerated vessel stirred by multiple-Rushton impellers, Chemical Engineering Journal 116, (2006), 85–95.
5. Nienow, A. W., Warmoeskerken, M. M. C. G.,Smith, J. M. and Konno, M., Flooding/loading transition and the complete dispersal condition in aerated vessels agitated by a Rushton turbine, *Proc 5th Europ Mixing Conf, Germany*, (BHRA Fluid Engineering, Cranfield) (1985),153-154.
6. Kudrewizki.F and Rabe. P, Model of the dissipation of mechanical energy in gassed stirred tanks, Chem. Eng. Sci., Vol. 41, No. 9, (1986), pp. 2247-2251.
7. Calderbank, P.H., 1958. The interfacial area in gas–liquid contacting with mechanical agitation. Transactions of the Institution of Chemical Engineers 36, 443–463.
8. Mounir Bouaifi and Michel Roustan, Power consumption, mixing time and homogenisation energy in dual-impeller agitated gas–liquid reactors, Chemical Engineering and Processing 40, (2001),87–95.
9. Lokhart, R.W., and Martinelli. R.C., Proposed correlation of data for isothermal two-phase, two component flow in pipes Chem.Eng.Progr. vol.45, (1949) 39-48
10. Wallis. G.B., One-dimensional Two-phase flow, McGraw-Hill, Inc. (1969)
11. Nicklin, D. J., "Two phase bubble flow", Chem. Eng. Sci., 17, (1962) 693-702.
12. Zuber, N., Findlay, J.A., Average voloume concentration in two-phase flow systems, J Heat Transf (Trans ASME), series C, 87, (1965), 453 – 468.
13. Hughmark, G. A.,Hold-up and mass transfer in bubble columns, Ind. Eng. Chem. Proc. Des. Dev., 6, (1967) 218-220.
14. Freedman, W. and Davidson, J. F., Hold-up and liquid circulation in bubble columns, Trans IChemE, 47: (1969) T251.

15 Viswanathan. K and D. Subba Rao, Circulation in bubble columns, Chem.Eng.Sci., 38 (1983)474-478.

16. Viswanathan K and D.Subba Rao, Inviscid liquid circulation in bubble columns Chem. Eng. Comm., 25 (1-6): (1984) 133-156.
17. Crabtree, J. R. and Bridgewater, J., Chain bubbling in viscous liquids, Chem Eng Sci, 24 (1969), 1755–1769.
18. Rietema, K. and Ottengraph, S. P. P., Laminar liquid circulation in bubble street formation in gas-liquid formation, Trans IChemE, (1970) 48: T54.
19. Heijnen, J.J., J. Hols, R.G.J.M. van der Lans, H.L.J.M. van Leeuwen, A. Mulder and R. Weltevrede, A Simple hydrodynamic model for the liquid circulation velocity in a full-scale two- and three-phase internal airlift reactor operating in the gas recirculation regime, Chem..Eng.Sci. , 52, (1997)2527-2540.
20. Gejork. K.M., Bujalski. W, Cooke. M and Nienow. A.W., The transition from homogeneous to heterogeneous flow in a gassed stirred vessel, Trans IChemE, vol 78, Part A, April (2000).
21. Sable. D, Hydrodynamics of gas-liquid system in a stirred tank, M.Tech Thesis, IIT Delhi, 1993.

An Investigation Into "Just Draw-Down" Conditions for Immiscible Liquid Systems

David A. R. Brown, BHR Group, Cranfield, United Kingdom

The "Just draw-down" condition is the minimum impeller speed required to ensure that a floating immiscible liquid later (in, for example, a liquid-liquid extraction process) is completely disturbed. Data obtained in a variety of stirred vessel geometries at a range of scales is presented, and performance prediction correlations are compared with those in the literature.

See more of this Session: Mixing in Multi-Phase Systems - Gas-Liquid and Liquid-Liquid Systems
See more of this Group/Topical: North American Mixing Forum

Prediction of Phase Inversion in Oil-Water Systems

Krzysztof Piela, Process Science Department, Unilever Food and Health Research Institute, Vlaardingen, Netherlands

Gijs Ooms, Laboratory for Aero- and Hydrodynamics, Delft University of Technology, Delft, Netherlands

Jan V. Sengers, University of Maryland, College Park, MD

Two immiscible liquids (for instance, oil and water) flowing at high velocity through a pipe mix due to turbulence may form a dispersed mixture, where one liquid is present in the other one in the form of drops. Which liquid is the continuous phase and which the dispersed one depends on many parameters, such as the viscosities and densities of the liquids, the presence of a surfactant, the 'history' of the experiment, the wetting properties of the pipe wall, etc. At a certain condition the continuous phase might become the dispersed one, and vice versa. This phenomenon is called phase inversion. Phase inversion has been studied for many years, but there are still many questions about it.

Two types of phase-inversion experiments are reported in the literature: continuous experiments and direct experiments. During a continuous experiment the dispersed phase is gradually added to the continuous phase and so the volume fraction of the dispersed phase increases with time. During a direct experiment the two liquids are mixed from the start at certain values of the phase-volume fraction for the two phases, which remain constant with time. We carried out continuous and direct experiments in a pipe (for details see our previous papers - Piela et al. (2006, 2008)). In our experiments we paid attention to the region between the maximum and minimum value of the dispersed phase-volume fraction where phase inversion can take place (for a certain set of conditions). During the continuous experiments we found that the critical volume fraction of the dispersed phase at the point of inversion does not depend on the Reynolds number, Froude number and Weber number of the mixture flow and also not on the injection velocity of the dispersed phase, as long as the mixture velocity is sufficiently large (> 2 m/s). However, we found that the injection phase-volume fraction (ratio of the rate of injection of the dispersed phase and the rate of the mixture flow through the pipe) had a significant influence on the critical volume fraction. Starting experiments with water flow and injecting oil leads to a different critical concentration than starting with oil flow and injecting water. The difference between these two is called ambivalence region. We observed that ambivalence region decreases with increasing injection phase-volume fraction. The direct experiment can be considered as the limiting case of a continuous experiment, with a large injection rate of the dispersed phase and with inversion taking place before one cycle through the pipe is completed. The injected phase-volume fraction is then equal to the ratio of the rate of injection of one of the two phases and the rate of the mixture flow through the pipe. The phase with the lowest injection rate is considered as the injected dispersed phase. Most of the experiments described in the literature, are in fact direct experiments where the influence of different parameters (viscosity ratio, density ratio, wetting properties, etc.) on inversion have been observed. Based on these observations a number of modelling approaches have been developed. In each of them a certain phenomenon plays a crucial role:

1) Coalescence. Under steady state conditions the break-up and coalescence of drops are in equilibrium. However, when the conditions change the coalescence rate may increases and exceed the break-up rate, leading to inversion (Yeo et al.(2002), Brauner et al.(2002)).

2) Formation of multiple drops. Small parts of the continuous phase are trapped inside the dispersed phase. So the drops (forming the dispersed phase) grow, coalesce and finally form the new continuous phase (Pacek et al (1994,1995), Pal (1993), Sajjadi et al (2000, 2002, 2003), Liu et al. (2005, 2006), Jahanzad et al. (2009)).

3) Dissipation rate. Inversion occurs when the effective viscosity of the oil-in-water mixture and the water-

in-oil mixture are the same (Poesio et al. (2008), Ngan et al (2009)).

None of the proposed models predicts phase inversion accurately. Models based on coalescence rate, formation of multiple drops and dissipation rate do not take into account 'history' of the experiment, which in many cases leads to wide ambivalence range. Ginzburg-Landau model can be used to describe ambivalence region in an oil-water mixture. This is due to the close similarity between the hysteresis region for first-order phase transitions and the ambivalence region for phase inversion between two liquids in a dispersed flow. According to our measurements, the ambivalence region is unique for a specific water-oil mixture. For such a mixture the ambivalence region depends on the injected phase-volume fraction. It is independent of the Reynolds, Froude, and Weber numbers and also independent of the injection rate, as long as the mixture velocity is large enough. Of course, when the oil type is changed or when a surfactant is added, the ambivalence region will change.

In many practical applications viscosity difference between oil and water phase is high. Reaching mixture velocity required to ensure that results are independent of the injection rate is in many applications impossible. At lower velocities, more viscous liquid will always tend to become dispersed phase. That of course has an impact on inversion map. In many experiments described in the literature, it was noted that increasing oil viscosity shifts inversion point towards lower oil volume fractions. We modified our Ginzburg-Landau model to extend it to systems with high viscosity ratio. In Ginzburg-Landau model phase inversion is predicted based on two parameters: injected phase volume fraction and friction factor. Friction factor represents energy dissipation. When two liquids have similar viscosities energy dissipation is almost the same for oil continuous and water continuous systems (system is symmetrical). Increasing oil viscosity leads to high increase in dissipation energy for oil continuous system, which affects the symmetry of the inversion map. Modification of the model to take into account high viscosity ratios allow us to extend model predictions to more viscous (practical) systems.

References

Brauner, N., Ullmann, A., 2002. Modeling of phase inversion phenomenon in two-phase pipe flows. International Journal of Multiphase Flow 28, 1177-1204.

Jahanzad, F., Crombie, G., Innes, R., Sajjadi, S., 2009. Catastrophic phase inversion via formation of multiple emulsions: A prerequisite for formation of fine emulsions. Chemical Engineering Research and Design 87 (4), 492 - 498, 13th European Conference on Mixing: New developments towards

more efficient and sustainable operations.

Liu, L., Matar, O. K., Lawrence, J., Hewitt, G. F., 2006. Laser-induced fluorescence (lif) studies of liquid-liquid flows. part i: Flow structures and phase inversion. Chemical Engineering Science 61, 4007-4021.

Liu, L., Matar, O. K., Perez de Ortiz, E. S., Hewitt, G. F., 2005. Experimental investigation of phase inversion in a stirred vessel using lif. Chemical Engineering Science 60, 85-94.

Ngan, K. H., Ioannou, K., Rhyne, L. D., Wang, W., Angeli, P., 2009. A methodology for predicting phase inversion during liquid-liquid dispersed pipeline flow. Chemical Engineering Research and Design 87 (3), 318 - 324.

Pacek, A.W., Nienow, A.W., 1995. A problem for the description of turbulent dispersed liquid-liquid systems. International Journal of Multiphase Flow 21, 323 - 328.

Pacek, A.W., Nienow, A.W., Moore, I. P. T., 1994. On the structure of turbulent liquid-liquid dispersed flows in an agitated vessel. Chemical Engineering Science 49 (20), 3485-3498.

Pal, R., 1993. Pipeline flow of unstable and surfactant-stabilized emuslions. AIChE Journal 39 (11), 1754-1764.

Piela, K., Delfos, R., Ooms, G., Westerweel, J., Oliemans, R. V. A., 2008. On the phase inversion process in an oil-water pipe flow. International Journal of Multiphase Flow 34, 665-677.

Piela, K., Delfos, R., Ooms, G., Westerweel, J., Oliemans, R. V. A., Mudde, R. F., 2006. Experimental investigation of phase inversion in an oil-water flow through a horizontal pipe loop. International Journal of Multiphase Flow 32, 1087-1099.

Piela, K., Ooms, G., Sengers, J. V., 2009. Phenomenological description of phase inversion. Physical Review E (Statistical, Nonlinear, and Soft Matter Physics) 79 (2), 021403.

Poesio, P., Beretta, G. P., 2008. Minimal dissipation rate approach to correlate phase inversion data. International Journal of Multiphase Flow 34 (7), 684-689.

Sajjadi, S., Jahanzad, F., Yianneskis, M., Brooks, B.W., 2003. Phase inversion in abnormal o/w/o Emulsion. 2. effect of surfactant hydrophilic-lipophilic balance. Ind. Eng. Chem. Res. 42, 3571 - 3577.

Sajjadi, S., Zerfa, M., Brooks, B. W., 2000. Morphological change in drop

structure with time for abnormal polymer/water/surfactant dispersions. Langmuir 16, 10015-10019.

Sajjadi, S., Zerfa, M., Brooks, B. W., 2002. Dynamic behaviour of drops in

oil/water/oil dispersions. Chemical Engineering Science 57, 663 - 675.

Yeo, L. Y., Matar, O. K., Perez de Ortiz, E. S., 2002. A simple predictive tool for modelling phase inversion in liquid-liquid dispersions. Chemical Engineering Science 57, 1069-1072.

See more of this Session: Mixing in Multi-Phase Systems - Gas-Liquid and Liquid-Liquid Systems
See more of this Group/Topical: North American Mixing Forum

Fabrication of Monodisperse Microcapsules with Stimuli-Responsive Membranes in Microfluidic Devices

Liang-Yin Chu, School of Chemical Engineering, Sichuan University, Chengdu, Sichuan, China

Abstract

Here we report a novel and simple method for preparation of monodisperse hollow microcapsules with pH-responsive membranes by combining microfluidics and interfacial cross-linking reaction. Using a capillary microfluidic device, uniform-sized oil-in-water-in-oil (O/W/O) emulsions, in which the water phase contains water-soluble chitosan and the inner oil phase contains terephthalaldehyde, are generated as templates for synthesizing microcapsules. The O/W/O emulsions were converted into microcapsules via the cross-linking reaction between terephthalaldehyde molecules and chitosan chains at the O/W interface. The prepared microcapsules have good monodispersity and sphericity with uniform hollow cavity and dense membrane structure, and exhibit diverse pH-sensitive characteristics. In acidic medium with pH ranging from 1.5 to 4.7, the microcapsules swell first, then collapse under osmotic pressure, and finally decompose and dissolve in the acidic solution. On the other hand, in weakly acidic or neutral medium with pH ranging from 5.8 to 7.1, the microcapsules only shrink and finally reach equilibrated state retaining good sphericity and structural integrity.

Keywords: Microcapsules; pH-response; Microfluidics

See more of this Session: Mixing in Multi-Phase Systems - Gas-Liquid and Liquid-Liquid Systems
See more of this Group/Topical: North American Mixing Forum

Design Mixers to Minimize Erosion and Corrosion-Erosion

by

Julian Fasano (Mixer Engineering Co.)
Eric Janz (Chemineer, Inc.)
Kevin Myers (University of Dayton)

AIChE Annual Meeting
November 8 – 13, 2009
Nashville, TN,
Paper 211a

Introduction

There are numerous applications of mixers that deal with erosive solids, especially in the minerals processing and power industries. In many of these applications there is an erosion-corrosion synergistic effect on the wear of a mixer's wetted parts, particularly the impeller.

There are a large number of factors that can affect the rate of erosion. Many of these factors have been known and studied to some extent:

- Chemical environment
- Hardness of solids
- Density of solids
- Difference in liquid and solid density
- Percent solids
- Shape of solids
- Fluid regime (turbulent, transitional or laminar)
- Fluid rheology (e.g. pseudoplasticity)
- Hardness of the mixer's wetted parts
- Young's modulus of the mixer's wetted parts
- Hydraulic efficiency of the impeller (kinetic energy dissipation rates near the impeller blades)
- Impact velocity
- Impact frequency
- Angle of impact

Theoretically the rate of volume loss of material is due to the kinetic energy lost when a particle impacts a material.[1] This would suggest a velocity exponent of 2. However, presented below, experimental velocity exponents have ranged from 1.5 to 4.0 The general form of the equation relating erosion rate to velocity is given by:

$$E = K \cdot V^n f(\theta)$$

where E = *volumetric erosion rate*

K = *constant (function of all parameters other than* V *or* θ*)*

V = *particle velocity or relative velocity for rotating systems (impellers)*

n = *velocity exponent (can also be a function of other parameters)*

θ = *impingement angle*

Most investigators have used this general equation form.

Sapate and RamaRao[2] used a power law correlation between volumetric erosion rate and impingement velocity in a non-rotating system. They observed exponents on velocity of 1.91 to 2.52. The velocity exponent showed an increasing trend with increasing hardness of the alloys irrespective of the hardness of the erodent particles and the impingement angle of the alloys investigated.

Stack[3] and others investigated the effect that corrosion plays in an erosion environment. These investigators studied various parameters of the corrosion-erosion environment in a non-rotating system. They observed velocity exponents that ranged from 1.4 to 3.5. They concluded that exponents derived for erosion of alloys under erosion-dominated conditions can be correlated to those derived for the strictly ductile erosion process. These are typically very near the theoretical "2" for the strictly ductile erosion process. However, those for the erosion-corrosion dominated regime are higher than for the erosion dominated regime and were in the 2.5 to 3.5 range. A publication by the Hydraulics Institute[4] suggests that the erosion velocity exponent for pumps in slurry transport is on the order of 2.5-3.04.

Fort[5] and others studied pitched blade impellers 100 mm in diameter with a blade width of 20 mm in water-solid slurries under turbulent conditions. These impellers were studied at pitch angles of 20°, 35° and 45°. These studies included a slurry of 18.3% by volume of gypsum having a mean particle diameter of 0.1 mm and a 10% by volume slurry of 0.4 mm mean diameter sand particles. From their studies they concluded the following:

- Particles of the lower hardness gypsum generated uniform sheet erosion over the entire surface of the impeller while the particles of sand, having a higher hardness, generated predominately erosion of the leading edges of the impeller blade.
- The higher the hardness of the blade material, the lower the wear rate of the blade.
- The wear rate of the leading edge was not a function of pitch angle.
- Sheet erosion of the blades exhibits a maximum erosion rate between 20° and 45°.

Zheng[6] and others studied the erosion-corrosion synergistic effect in an acidic slurry. The slurry was 10% by weight H_2SO_4 and 15% by weight –60 mesh corundum sand. Their apparatus was a rotating disk with four specimen holders on its edge. They determined the rate of erosion by making electrochemical measurements during rotation. All of their studies were done under turbulent flow conditions. Erosion rate velocity exponents ranged from 1.9 to 4.0. A model was proposed and used which divided the overall erosion rate into an erosion rate via corrosion, an erosion rate via erosion, and an erosion rate due to synergism. The synergism rate was very large and varied between 32 and 99% of the total. The percent contributed by synergism diminished as the alloy became more statically corrosion resistant.

Amelyushkin and Agafonov[7] studied the erosion of cogeneration steam turbine blades caused by water droplets. If kinetic energy is high enough, even water droplets can cause erosion. They found that the toroidal and near root vortices were very intense and caused enhanced wear of the rotor blades. Also they found that they were able to eliminate erosion by making the water droplets small enough. It is expected that these effects are related to grain size. In ductile erosion, plastic deformation may occur first, before metal is removed. If erosion is due to inter-granular grain fracture, then if particles are significantly smaller than the metal's grain size, erosion should be minimized. As ductile alloy grain sizes are on the order of 20 μm, particles smaller than this should have little erosive effect.

Khalid and Sapuan[8] studied wear for a centrifugal pump impeller in a slurry application. Weight and diameter losses were very nearly linear with time over 480 hours of operation. Blade height and depth loss did exhibit some non-linearity but were modeled as linear. Typical of rotating devices in slurries, more material was lost near the periphery of the impeller than in the center because linear velocities increase with radial distance from the shaft.

Lopez[9] and others studied the effect of corrosion-erosion at relatively high velocities on 304 and 420 stainless steel. Velocities ranged from 4.5 to 8.5 m/s. Such velocities are not common in mixing equipment except in high shear devices. They used a rotating disc device with erosion samples attached to the periphery of the disc. The aqueous liquid for the slurry was composed of 70% by weight H_2SO_4 and 3.5% NaCl. The slurry solids were 30% by weight SiO_2 particles with a mean diameter about 0.25 mm. They found that high velocity impacts were beneficial. The combined action of erosive and corrosive mechanisms did not lead to a significant increase in mass loss if compared to corrosion tests. They suggested that pores, small cracks and fresh pits can be covered by the prows and lips that are formed as a consequence of the wedge action of round particles which produces a smoother, uniformly corroded surface. Thus even though the transport mechanisms which remove corrosion products are greater due to higher velocities, the surface area exposed is smaller.

Fasano and Corpstein[10] studied slurry wear through multilayer paint modeling. The paint layers are on the order of 0.0015 inches thick. The three layers of paint used had an overall thickness of approximately 0.0045 inches. This is only about 7% of the blade thickness and did not change the fluid hydraulics over the blade. Erosion was studied using 8.3 inch diameter scaled down axial flow mixing impellers in a sand water slurry. These studies pointed out the strong effects that blade-shedding vortices could have on erosive wear. Impellers, such as the four bladed-pitched impeller that generated stronger vortices suffered the highest degree of localized erosion. The effects of these vortices can completely wear through an impeller blade, leaving holes where the vortices contacted the surface. High efficiency impeller blades created significantly smaller vortices and as a consequence exhibited much lower localized wear. Vortex erosion can be severe and occurs on the backside (low pressure side) of the impeller blade. Comparisons of the backside wear pattern for a Chemineer HE-3 impeller and a standard generic 45° pitched four-bladed impeller are shown in Figures 1 and 2. The impellers were each 8.3 inches diameter and operated at 870 rpm in a 10% by weight sand slurry in water. The weight mean particle size of the sand was 360 μm. These backside erosion patterns were made after only 30 minutes of operation and it is obvious that the erosion on the backside of the 45° four-bladed pitched impeller was much more severe than the erosion for the Chemineer HE-3 high efficiency impeller.

Figure 1: Backside of Pitched Blade Impeller

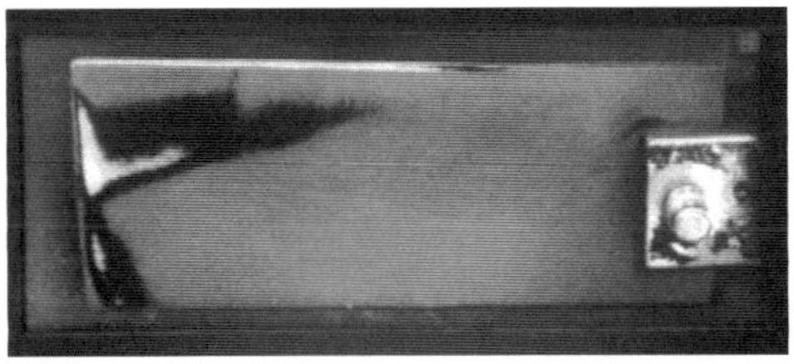

Figure 2: Backside of HE-3 High Eff. Impeller

Wu[11] and others were also successful in using this technique to study five different style radial flow impellers and a low attack angle (~15°) (6-bladed) pitched impeller. As expected, the hydraulically more efficient six bladed-pitched impeller experienced the least erosion.

Increased hardness of metals will generally provide an increased life. Miller and Schmidt[12] compared the erosion rates of 16 metals in a recycled slurry test system using 2% by weight silica sand in water. The impeller tip velocity was 15.7 m/s and the temperature was 16°C. In addition to the erosion rate for each metal, they included the metal's hardness. The best fit for their data was logarithmic. However, probably due to the synergistic corrosion effects, the data was fairly dispersed. A plot of this data is provided in Figure 3. The effect of particle hardness depends on whether erosion is ductile or brittle. For brittle erosion, the effect of particle hardness is much more pronounced than for ductile erosion.

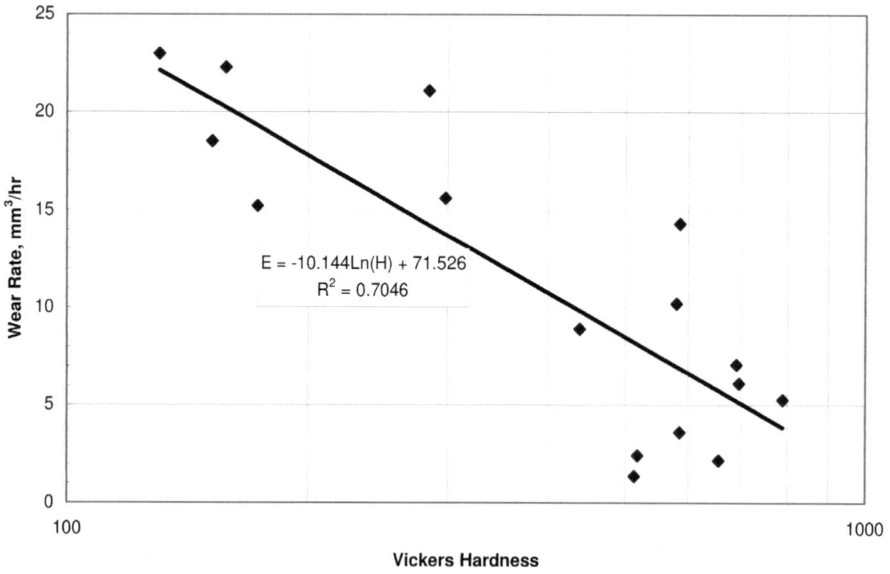

Figure 3: Wear Rate Data of Miller and Schmidt

Changes in particle size can change the erosion mechanism. Stachowiak and Batchelor[13] reported that as the particle size was increased from 8.75 μm to 127 μm, the mode of erosion changed from ductile to brittle. The erosion study was for silicon carbide particles impinging on glass, steel, graphite and ceramics. The particle velocity was 152 m/s.

Design for Erosion Minimization

Because maximum velocities in mixing processes seldom exceed 6 m/s, erosion and corrosion-erosion of materials are fatigue processes for most mixing processes. There's generally not enough particle kinetic energy to cause ductile erosion where there is some plastic flow of material. The fatigue process occurs on a micro or localized scale, and, as with macro-scale fatigue, two stages of the erosion process have been observed. There is an incubation period followed by the formation and growth of pits involving the removal of the metal or material. Refer to a materials behavior text such as that by Hertzberg[14] for a more in-depth discussion on material behavior.

Due to the vast number of parameters that can affect erosion or erosion-corrosion processes, and the fact that this area of mixer service has not been widely studied, it is very difficult to predict a priori what the rate of erosion will be for any given liquid-solid application. However, there are certain factors within the control of the designer that can be used to optimize the life of the mixer's wetted parts.

Most mixer designers will not have control over the type of slurry, the percent solids, the hardness of the solids, the shape of the solids, the liquid, the pH, etc. However designers will generally have control over:

- The mixer wetted parts materials, coating or lining
- The impeller style
- The impeller horsepower and speed combination

Material Selection

The choice in selecting a material is to either go hard or soft and elastic. All else being equal, when selecting a metal alloy, a higher hardness will lead to a longer life. Thus when selecting a metal alloy material, select a hard material which will also provide good corrosion resistance.

There are a number of hard surface ceramic coatings such as tungsten carbide or silicon carbide, which could be applied to the high-wear areas such as impeller blades. Ceramics are the most wear resistant but are low in toughness and impact strength. Ceramic coatings as well must be corrosion resistant to the liquid medium. Ceramics also do not have the ability to absorb much strain. So high strains on flexing blades may allow cracking of the ceramic coating. Ceramic coating applicators should be able to provide the maximum allowable strain for the ceramic coating under consideration. Coatings of the more common ceramic materials tend to be more costly than high hardness metals, or elastomeric coverings.[15]

Glass lined equipment has a glass hardness of 5 to 6 on the moh scale. For the great majority of solids, this hardness would be very acceptable. However there are numerous materials and minerals including: Al_2O_3, SiO_2, WC, SiC, and ZiO_2 that have higher harnesses and would tend to wear away the glass lining. Glass linings have very many of the same limitations as ceramic coatings. They tend to be brittle and cannot tolerate much strain.

Elastomeric coverings on the order of 3/8" thick for industrial scale impellers have a long history of providing longer life in slurry applications. Instead of having to absorb most of the particle's impact energy, an elastomer releases most of the energy back to the particle after impact. Elastomeric lining manufacturers and applicators will generally recommend an elastomeric hardness of 40-60 Durometer A for optimum life. As with metals or hard surface coatings, the lining must also be compatible with the fluid medium. An elastomer's hardness is directly related to its corrosion resistance. However, as an elastomer's hardness increases above a 40A Durometer, its erosion resistance decreases. A Durometer selection of 40-60 A is somewhat of a compromise between erosion and corrosion resistance. Elastomers should not be used when large particles are present. The term "large particles" is relative to the impinging velocity and mass of the particle, as well as the thickness of the elastomeric covering. If the impinging particle can bottom out against the metallic substrate, elastomeric coverings should not be used. Even if most of the slurry might be suitable, a small percentage of tramp particles can do significant damage to the elastomeric covering. Since impact energy is a function of the impingement angle, the leading edges of impeller blades are almost always double wrapped. The most popular elastomeric coverings are: natural rubber, neoprene, butyl, chlorobutyl, and hypalon. Improperly applied linings on high-efficiency impellers that significantly change the profile of the blade can cause increased erosion problems. Typically linings are double layered on the leading, trailing and outside edges of impeller blades. These linings must be adequately feathered such that the transition from the double layer to the single layer is smooth to avoid generation of additional vortices.

Both thermoplastic and thermoset polymers do not have the ability to restore back to the particle most of the kinetic energy and are generally not as good in mixing slurry service. Hercules 1900 UHMWPE, touted as being a very abrasion resistant polymer, was tested by the authors against Polymeric Protective Coatings elastomers 2001-B (natural rubber Durometer A 30-40) and 1054-B (chlorobutyl rubber Durometer A 35-45). The testing procedure was identical to that specified in the Hercules 1900 UHMWPE bulletin. A 50% sand-water slurry was used with a sand weight mean particle size of 53 μm. The specimen tip speed was 2.22 m/s. Weight loss was determined at various time intervals over an 8 hour period. The weight loss versus time was found to be linear with R^2 values for all three falling between 0.93 and 0.94. The rates of erosion were:

Material	Rate of Weight Loss, g/hr
Hercules 1900 UHMWPE	0.0975
Polymeric Protective Coatings 2001-B (natural rubber)	0.0104
Polymeric Protective Coatings 1054-B (chlorobutyl)	0.0124

As can be observed, the rate of weight loss for the thermoplastic polymer is 7 to 10 times greater than that for the elastomers tested.

Dickey and Fasano have provided a general reference on materials selection considerations.[16]

Impeller Selection

There are many different impeller styles available to the designer. Selecting the correct impeller can often make a difference in impeller life of two or three times. Erosion of impeller blades can depend heavily on the flow regime, with flow regime being determined by the impeller Reynolds number.

$$N_{\mathrm{Re}} = \frac{\rho N D^2}{\mu}$$

$$\begin{aligned} \textit{where } \rho &= \textit{density} \\ N &= \textit{impeller rotational speed} \\ D &= \textit{impeller diameter} \\ \mu &= \textit{viscosity} \\ N_{\mathrm{Re}} &= \textit{Reynolds Number, dimensionless} \end{aligned}$$

Impellers in turbulent flow create shedding vortices that attach themselves to the back of impeller blades. There are a number of techniques that can be used to visualize these vortices. In Figure 4 and Figure 5 telltales attached to the blade are used to visualize these vortices.[17] They are shown for both a relatively inefficient 45° pitched four-blade impeller and the Chemineer HE-3 high efficiency impeller.

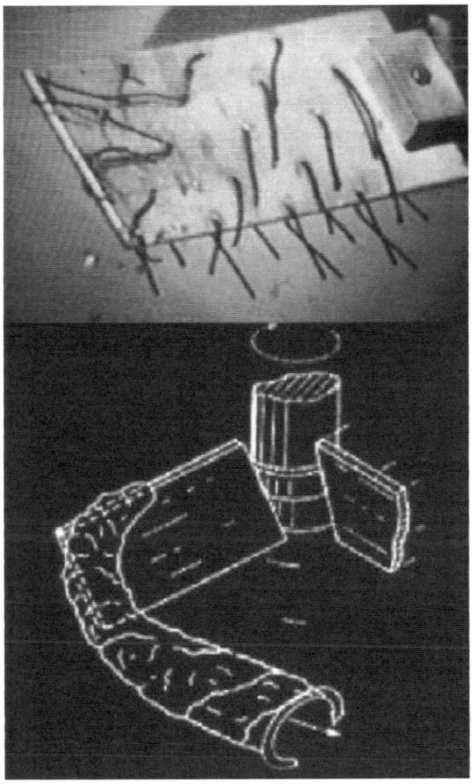

Figure 4: Pitched Blade Impeller Vortex

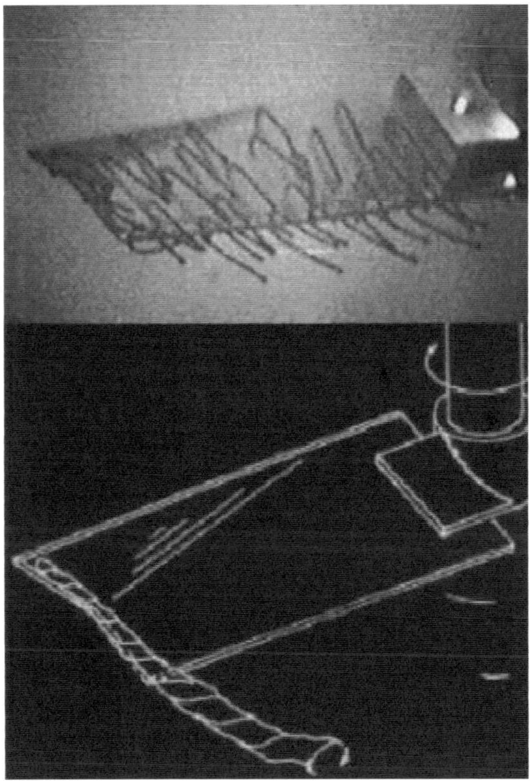

Figure 5: HE-3 Impeller Vortex

In these figures, the impellers are rotating clockwise when viewed from above. Thus the blades are moving into the page, and the view is of the backside of the blades. These vortices cause very localized wear emanating from the back-side or low pressure side of impeller blades. These impeller vortices begin to diminish at Reynolds numbers below 10,000, become very weak below a Reynolds number of 500, and have completely disappeared below a Reynolds number of 10. As most slurry particles are heavier than the fluid, the centrifugal effect caused by the vortex will cause particles caught in the vortex to migrate to the OD of the vortex. Thus the concentration of solids at the periphery of the vortex is much higher and the rate of solid particle to surface impacts is much greater, increasing locally the rate of erosion. Of course, particles must be small enough to be captured in these vortices before this effect would be observed. On an industrial scale, however, the greatest majority of slurry applications would have particles sufficiently small to be captured by these vortices.

There are a number of relatively efficient wide blade impellers used in solids suspension service. Fasano and Reeder[18] compared the erosion rate between a Chemineer Maxflo W impeller, Figure 6, and a standard 45° four-bladed pitched impeller (refer to Figure 4).

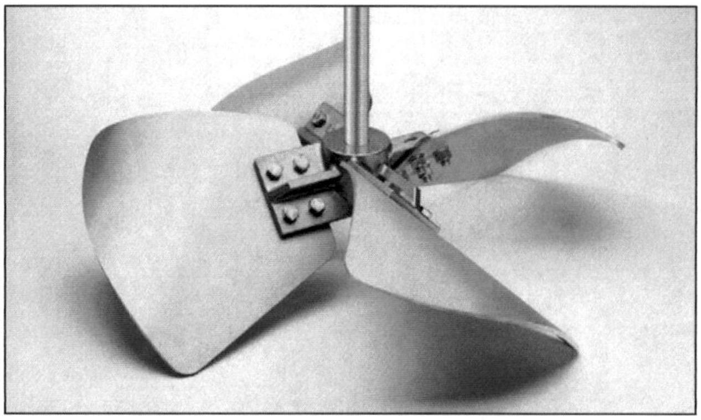

Figure 6: Chemineer Maxflo W Impeller

For the same level of solid suspension, these impellers utilize the same impeller diameter at the same speed. Therefore velocities at the impeller are the same. The rate of erosion however for the pitched-blade impeller was, on a percentage basis, 59% greater than the erosion rate of the Maxflo W impeller.

Radial flow impellers are not very efficient in suspending solids. However, radial flow impellers are efficient in dispersing gasses. In applications where solids are present and a gas must be dispersed they are often used. As with axial flow impellers, impeller efficiency changes with design. The Rushton or D-6 impeller, introduced in the late 1940's, creates a pair of vortices behind each blade in turbulent flow. These vortices, as we've observed with the pitched-blade impeller, can cause severe erosion from the backside of the blade. The Figure 7 photo of the intertwined telltales on the backside of the blade demonstrates the size and nature of these vortices.

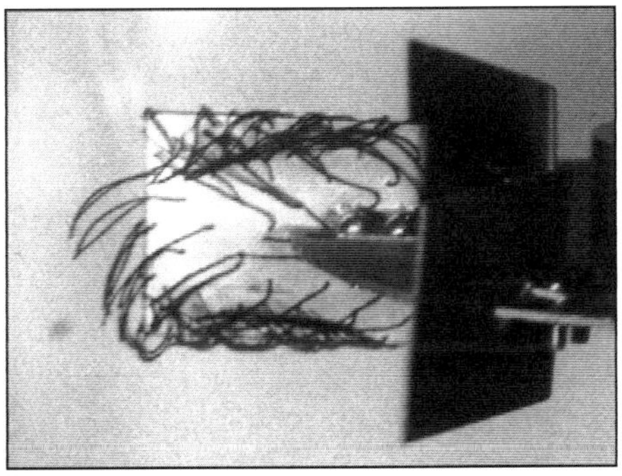

Figure 7: Telltales on D-6 Impeller Blade

There exist today highly efficient radial flow impellers. The Chemineer CD-6 impeller was introduced into the marketplace in 1988, and the even more efficient Chemineer BT-6 was introduced in 1998. Both of these impellers exhibit very little tendency for vortex formation on the back-side of the blade. This is demonstrated in telltale photos in Figure 8 and Figure 9. In each of these photos the blades are rotating into the plane of the paper.

Figure 8: Telltales on CD-6 Impeller Blade

Figure 9: Telltales on BT-6 Impeller Blade

Under turbulent flow conditions, it is usually very beneficial, both from an erosion standpoint as well as a process efficiency standpoint, to select a high efficiency style impeller. In transitional flow there are still benefits to be achieved by the use of a high efficiency impeller, but not as profound as in turbulent flow. In laminar flow, there is no advantage in using a high-efficiency impeller either for erosion or solids suspension efficiency.

Horsepower and Speed Selection

As discussed earlier, erosion is very dependent on velocity, and typically in erosion-corrosion environments, the velocity exponent for the volumetric rate or weight rate of erosion is generally observed to be 2.5 to 4.0. In designing an agitator for the suspension of solids, the designer has a choice of selecting a number of power and speed combinations. Because solid suspension impeller efficiencies change with impeller style, impeller to tank diameter and off-bottom clearance to tank diameter[19], a number of possible horsepower and speed choices can meet process objectives. The selection of a specific agitator design in the end should come down to economics. There are capital costs and operating costs. Capital costs are largely associated with the general size of the machine. The torque can generally best characterize the capital cost. Operating costs include the energy costs to operate the machine plus any maintenance costs. Maintenance costs include the costs of oil changes, new bearings, new gears, new seal components, and in the case of erosion applications, the replacement of in-tank wear components, typically impeller blades. A close examination of the various horsepower and speed options should be examined closely in order to make an economic selection.

Example

For the sake of demonstration let us assume that we need to design an agitator for suspending a 10% solution of sand in water. The sand will be assumed to have a weight mean particle size of 360 μm. The tank is 12 feet diameter with a 2:1 elliptical dished bottom, and the water-sand slurry will have a liquid volume such that the depth of liquid in the tank is 12 feet. The sand will be assumed to have a specific gravity of 2.4 and the water a specific gravity of 1.0. The viscosity of water will be assumed to be 1 mPa-s. The process solution requires that the solids be suspended to the "just suspended" condition such that no particles settle on the bottom of the tank for more than 2 seconds. All designs are to utilize a single HE-3 impeller one-third the tank diameter off-bottom. The horsepower, speed and impeller diameter combinations that satisfy the process ovjective can be determined using solids suspension design procedures such as the article presented by Corpstein and others.[20] The following designs shown in Table 1 below all satisfy the off-bottom solids suspension process requirement.

Table 1: Possible Process Design Selections for Example

Impeller Diameter, in	Impeller Dia./Tank Dia.	Impeller Speed, rpm	Impeller Power, hp	Impeller Torque, in-lb	Impeller Tip Speed, ft/min	Relative Wear Life *	Approx. Relative Cap. Cost
35	0.243	159	4.68	1855	1457	1.00	1.00
40	0.278	123	3.94	2019	1288	1.54	1.06
45	0.313	102	4.22	2607	1202	1.96	1.25
50	0.347	92	4.81	3295	1204	1.95	1.45
55	0.382	84	5.43	4074	1210	1.92	1.67
60	0.417	75	6.18	5193	1178	2.10	1.95
65	0.451	70	7.01	6311	1191	2.02	2.22
70	0.486	67	9.15	8607	1228	1.82	2.71
75	0.521	65	12.28	11907	1276	1.59	3.35

* Relative wear based on assumed velocity exponent of 3.5

Even though using small impellers operating at high speeds reduces capital costs and power, the high tip speeds of these designs lead to short wear lives. Using an intermediate size impeller that minimizes tip speed at the cost of higher capital and power costs maximizes the wear life. As can be observed, the torque, and consequently the capital cost, increase dramatically with impeller to tank diameter ratios greater than 0.45 due to changes in the flow pattern generated by the impeller.

Scale-Down Studies

Since there is often an erosion-corrosion synergistic effect that cannot be predicted a priori with today's current data, any material selection should be studied on a smaller scale, before making a final choice. Since it is important to model the same hydraulic behavior over the blade, the authors recommend that the ratio of mean particle diameter to impeller diameter not exceed 0.008. For example, a mean particle diameter of 1 mm would suggest a small-scale impeller no less than 125 mm (4.9 in) diameter. It is also important to ensure that fluid regimes haven't changed. If the impeller operation is turbulent on the large scale, it should also be turbulent on the small scale.

An optimum agitator horsepower and speed selection can be made as described above for the full scale. However, in order to determine the rate of erosion, scale-down studies should be made. A geometric scale-down for an equal level of solid suspension will result in a tip speed that will always be lower on the smaller scale, except when scaling down geometrically for very slow settling solids (<0.1 m/min). Wear rate, as previously demonstrated, is a strong function of velocity. Therefore all scale-down tests should be made at equal tip speed. As an example, if we examine the 55-inch diameter impeller solution above for the above described problem and scale this down to an 18-inch diameter tank we would have the following comparison.

Table 2: Comparison of Equal Suspension to Equal Tip Speed

Condition	Impeller Dia., inches	Impeller Speed, rpm	Impeller Tip Speed, ft/min
Full Scale Design, 144 in. dia. Tank	55.0	84	1210
Scale-down Design based on Equal Suspension	6.875	456	821
Scale-down Design Based on Equal Tip Speed	6.875	672	1210

Conclusions and Summary

The rate of erosion is dependent on the following major static environmental factors: chemical environment, hardness of the solid particles, density of the solid particles, percent solids, the shape of the solids, the size of the solids including whether or not tramp solids are present, and type of impeller. The dynamic factors affecting erosion rate are: fluid regime, impact velocity, impact frequency and angle of impact. As there are no good means currently of predicting erosion rate, small-scale studies should be conducted emulating as much of the total environment as possible. These small-scale studies should be conducted using equal tip speed to mimic the full-scale rate of erosion.

Erosion in most mixing processes is a fatigue process generally accelerated by a liquid corrosive environment. The fatigue process occurs on a micro or localized scale, and, as with macro-scale fatigue, two stages of the erosion process have been observed. There is an incubation period followed by the formation and growth of pits involving the removal of the metal or material. One of two routes is generally utilized in dealing with an erosion application. The high velocity areas such as the blades are either made from hard materials, or coated with hard ceramic materials such as tungsten carbide or silicon carbide. Alternatively, the blades are covered with some type of elastomeric covering.

Impeller selection is important especially in turbulent flow conditions. High efficiency impellers will generally erode at a slower rate because the backside of the blades has minimized shedding vortices. In laminar flow, from an erosion standpoint, most impellers behave similarly due to a lack of vortices. Thus, the selection of the impeller should be based primarily on what is required to accomplish the desired process result.

A number of horsepower and speed selections that satisfy the process requirement should be examined to conduct an economic analysis. The lowest possible speed may not be the most economical. It is best to first design the most cost optimum agitator for the full scale. Then in order to estimate the corrosion rate, scale down on the basis of equal tip speed.

References

1. Chattopadhyay, R., **Surface Wear, Analysis, Treatment and Prevention**, ASM International, Metals Park, OH, 2001, pp. 80-90.
2. Sapate, S. G. and RamaRao, A. V., **Effect of Erodent Particle Hardness on Velocity Exponent in Erosion of Steels and Cast Irons**, Materials and Manufacturing Processes, Vol. 18, Issue 5, Jan. 2003, pp. 783-802.
3. Stack, M. M., Stott, F. H., and Wood, G. C., **The Significance of Velocity Exponents in Identifying Erosion-Corrosion Mechanisms**, Journal de Physique IV, Colloque C9, Supplement au Journal de Physique III, Volume 3, Dec. 1993.
4. **A New Slurry Pump Standard**, Pumps and Systems, The Hydraulic Institute, March 2006, pp. 66-69.
5. Fort, I, Medek, J, Ambros, F., **Erosion Wear of Axial Flow Impellers in a Solid-Liquid Suspension**, Acta Polytechnica, Vol. 41, No. 1/2001, Czech Technical University Publishing House, pp. 23-28.
6. Zheng, Y., Yao, Z., Wei, X., Ke, W., **The Synergistic Effect Between Erosion and Corrosion in Acidic Slurry Medium**, Wear 186-187, 1995, pp. 555-561.
7. Amelyushkin, V. N., and Agafonov, B. N., **Special Features of Erosion Wear of Rotor Blades on Cogeneration Steam Turbines**, Power Technology and Engineering, Vol. 36, No. 6, 2002, pp. 359-362.
8. Khalid, Y. A., Sapuan, S. M., **Wear Analysis of Centrifugal Slurry Pump Impellers**, Industrial Lubrication and Tribology, 59/1, 2007, pp. 18-28.
9. Lopez, D., Congote, J. P., Cano, J. R., Toro, A., and Tschiptschin, A. P., **Effect of Particle Velocity and Impact Angle on the Corrosion-Erosion of AISI 304 and AISI 420 Stainless Steel.**, Wear, 259, 2005, pp.118-124.
10. Corpstein, R. C., Fasano, J. B., **Erosion of Rubber Covered Impeller Blades in an Abrasive Service**, The Indian Chemical Engineer, Vol. 36, No. 1, 1990.
11. Wu, J., Ngyuen, B., Graham, L., Shu, Y., Kilpatrick, T., and Davis, J., **Minimizing Impeller Slurry Wear Through Multilayer Paint Modeling**, The Canadian Journal of Chemical Engineering, Vol. 83, October 2005, pp. 835-842.
12. Miller, J. E., Schmidt, F., **Slurry Erosion: Uses, Applications and Test Methods**, ASTM Publication 04-946000-29, Published by ASTM International, 1987, p. 177.
13. Stachowiak, G. W., Batchelor, A. W., **Engineering Tribology**, 3rd Edition, Butterworth-Heinemann, 2005, pp. 530-533.
14. Hertzberg, R. W., **Deformation and Fracture Mechanics of Engineering Materials**, 4th Ed., J. Wiley & Sons, New York, 1995.
15. Wilson, K.C., Addie, G.R., Sellgren, A., Clift R., **Slurry Transport Using Centrifugal Pumps**, 2nd Ed, Springer Publishing, 2005, p. 234.
16. Dickey, D. S., and Fasano, J. B., **Handbook of Industrial Mixing**, Edited By Paul, Atiemo-Obeng and Kresta, Chapter 21, Section 9, John Wiley and Sons, Inc. New Jersey, 2004.
17. Fasano, J. B., **Flow Visualization Techniques on Rotating Impellers**, Engineering Foundation Mixing Conference XII, Pitosi, Missouri, August 6, 1989.
18. Fasano, J. B., Reeder, M. F., **An Improved Maxflo Impeller**, North American Mixing Forum, Mixing Conference XVII, Banff, Alberta, Canada, August 15-20, 1999, Paper 2.4.
19. Myers, K.J., Corpstein, R.R., Bakker, A., Fasano, J.B., **Solid Suspension Agitator Design with Pitched Blade and High Efficiency Impellers,** AIChE 1993 Annual Meeting, November 7-12, 1993, St. Louis, Missouri, Paper No. 1c.
20. Corpstein, R. R., Fasano, J. B., Myers, K. J., **The High-Efficiency Road to Liquid-Solid Agitation**, Chemical Engineering, October 1994, pp. 138-144.

SUCCESSES AND EMERGING ISSUES IN SIMULATING THE MIXING BEHAVIOR OF LIQUID-PARTICLE NUCLEAR WASTE SLURRIES AT THE SAVANNAH RIVER SITE – 211b

David C. Koopman, Savannah River National Laboratory, Aiken, SC
Bradley R. Pickenheim, Savannah River National Laboratory, Aiken, SC
Daniel P. Lambert, Savannah River National Laboratory, Aiken, SC
Michael E. Stone, Savannah River National Laboratory, Aiken, SC

ABSTRACT

Aqueous radioactive high-level waste slurries are combined during processing steps that ultimately produce a stable borosilicate glass waste form. Chemically treated waste slurries are combined with each other and with glass frit-water slurries to produce the melter feed. Understanding the evolution of the rheological properties of the slurries is an important aspect of removing and treating the stored waste. To a first approximation, combinations of colloidal waste slurry with ~0.1-mm mean diameter glass frit or glass beads act in an analogous matter to slurries of spherical beads in Newtonian liquids. The non-Newtonian rheological properties of the waste slurries without frit, however, add complexity to the hydrodynamic analysis. The use of shear rate dependent apparent viscosities with the modified Einstein equation was used to model the rheological properties of aqueous frit-waste slurries.

INTRODUCTION

Legacy radioactive high-level waste (HLW) generated at the Savannah River Site (SRS) during Cold War production of enriched uranium and plutonium is currently being processed into a stable borosilicate glass waste form for long term storage. The majority of the legacy waste is stored as a mixture of hydroxide and hydrous oxide insoluble solids in large cylindrical storage tanks at SRS. Over 90% of the waste solids are non-radioactive chemical byproducts derived from the fuel rod targets and purification chemistry. The 3.5-4.9 thousand cubic meter (900,000-1,300,000 gallon) carbon steel waste storage tanks also contain 5-7M sodium solutions rich in hydroxide, nitrate, and nitrite anions that are contaminated with soluble radioactive isotopes of cesium and strontium. Insoluble solids have been allowed to settle to the bottom of the tanks and have been aging for 25-50 years. These solids must be mobilized and transferred between tanks as the first step in waste treatment.

The list below summarizes some of the tank farm issues:

- Rheological properties of suspended waste solids degrade with time under shear (the slurry becomes more viscous).
- Rheological properties need to be understood as a function of the insoluble solids content, particle size distribution, and composition.
- Slurry pump flow fields in the large, cooling coil filled, tanks have numerous stagnant zones where radioactive waste mounds form creating closure issues.
- Settling times during large tank washing operations are not predictable. Settling times also do not mimic those of small scale radioactive settling test samples.

- Radiolytic hydrogen generation and bubble accumulation during gravity settling constrains washing volumes.
- Addition of glass frit to the slurries prior to the waste glass melter produces a shift in the rheological character of the slurry.

SRNL has a program that produces non-radioactive "simulants" of the tank farm wastes in sufficient volumes, 50 to 5,000 dm^3 (liters), to study multi-phase mixing and transport issues without the radiation hazards of the real waste. Simulant slurries are used in small to intermediate scale equipment to evaluate the potential feasibility of various proposed process operations. Although simulant preparation chemistry generally follows that of the original radioactive wastes, the physical properties (rheology, particle size, cohesiveness, foaming tendency, etc.) often do not adequately match those of the actual wastes. The radioactive and simulant slurries are generally maintained at insoluble solids loadings where they behave like pseudo-homogeneous, thixotropic (shear thinning) fluids. Properties and appearance are similar to those of watered-down clay or the red mud in bauxite refining. Typical particles sizes for the precipitated solids are less than 40 micrometers, including a significant fraction of sub-micron sized particles.

Significant progress has been made in producing rheologically similar simulants. Testing has shown that batch precipitated simulants tend to be more viscous than those produced using continuous stirred tank crystallizers. High-shear mixers have been used to increase the apparent viscosity of simulants as necessary. Unfortunately, this can come at the expense of creating a mismatch in particle size distribution that may impact studies of other phenomena such as foaming or bubble retention. Actual waste rheological properties must be adjusted (by decanting or dilution) into certain ranges for successful tank mixing, solids suspension, and pump inter-tank transfers while retaining sufficient viscosity to minimize segregation of particles during anticipated worst case pump outage scenarios.

A fairly successful method for predicting the Bingham plastic yield stress and consistency of waste slurry blends was developed (Koopman, D., "Successes and Emerging Issues in Simulating the Processing Behavior of Nuclear Waste Blends", *Proceedings of the 2009 AIChE Annual Meeting*, New York, NY: AIChE. 2009). This method involves taking individual slurry rheometer data in the form of shear-rate dependent apparent viscosities, i.e. shear stress/shear rate, and applying empirical mixing rules for viscosity.

Properties of liquid-particle HLW slurries change significantly in the final processing step just prior to the waste melter. Glass frit is added at a mass equivalent to about 2-3 times the mass of the waste solids. The frit provides the chemical backbone that holds the HLW species in a stable glass waste form after processing through the melter. The frit particles are essentially fine shards of ground glass with particles in the 80-200 mesh range (0.07-0.20 mm) based on the Tyler standard screen scale. Figure 1 shows the essentially non-overlapping nature of the particle size distributions of the waste species and the glass frit. The frit peak is centered at about 150 microns, while the waste solids peak is at about two microns.

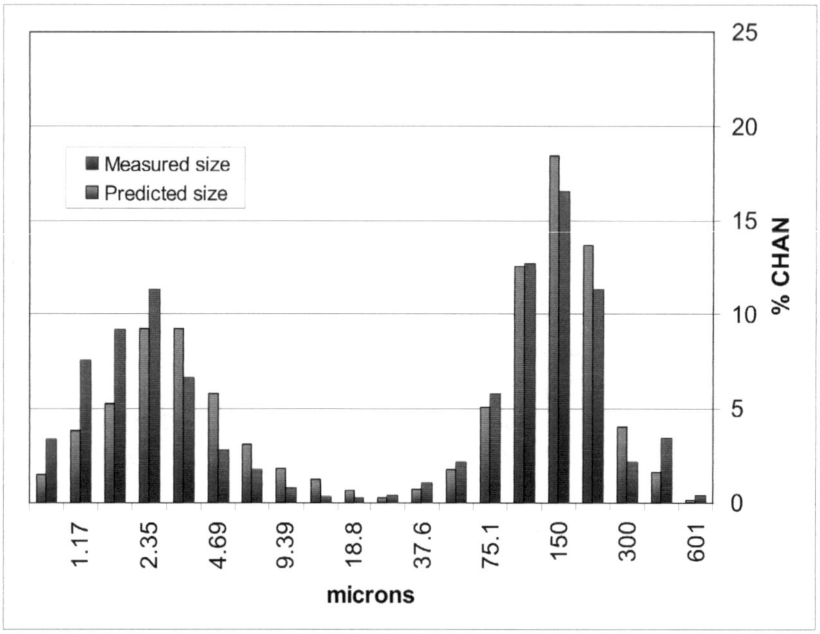

Figure 1. A comparison of the particle size distribution of melter feed with a weighted average of the individual particle size distributions for sludge and frit.

Frit particles are very abrasive, and it may be that the difference between measured and predicted particle size on the low end is partly due to attrition of the waste particles coming into contact with the frit particles under shear.

The impact of the frit particles on rheology is more comparable to adding 0.1 mm glass beads to water than to adding additional fine particles like the precipitated wastes. The bead-water system behaves somewhat like a single-phase Newtonian fluid in certain shear rate ranges even though it is clearly a two-phase medium, that is, the removal of shear would lead immediately to rapid segregation of the beads from the bulk of the water. The viscosity of the bead-water suspension increases with increasing volume fraction of beads. Various models have been proposed, such as the extended Einstein equation. As the volume fraction of beads increases, the models become progressively more complex. Modeling the flow of bead-water mixtures as a single-phase flow subject to certain constraints offers certain advantages over using the equations developed for heterogeneous flows.

A fairly viscous HLW slurry is required to ensure that glass frit particles do not rapidly segregate relative to the waste particles during mixer outages and inter-tank transfers. SRS contractors are required by the U.S. Department of Energy to verify that they are producing a stable homogeneous melter feed slurry with acceptable waste form properties so that direct sampling of the poured glass stream is not required.

EXPERIMENTAL AND MODELING WORK

Progress has been made on modeling the melter feed rheology as a pseudo-homogeneous Bingham plastic fluid (based on the near-colloidal sized fine waste solids) in two-phase flow with the larger frit particles using a variation of the extended Einstein equation. The rheological properties of the frit-free slurry are obtained in the form of an apparent viscosity, μ, (as a function of shear rate, γ) and used with the volume fraction, φ, of the frit particles to calculate the shift in viscosity from the entrained large particles:

$$\mu_{melter\ feed} = \mu_{sludge} * (1 + 2.5 * f * \phi + 10 * \phi^2)$$

Frit can occupy 7-14 volume percent of the melter feed slurry. The factor, f, was introduced to the extended Einstein equation to account for deviations from spherical particles. The constant of ten on φ^2 is also semi-empirical (and there are other values of this constant proposed by other researchers, e.g. Guth and Simha proposed 14.1 in 1936). The equation was developed for spheres in a Newtonian fluid, and the constant on the quadratic term was not adjusted during this study. A four term expansion has also been proposed containing the above terms plus an exponential term (Thomas, 1965). The primary use for this equation is in predicting the impact of changing the ratio of frit to waste solids on rheological properties without having to repeat series of rheometer measurements on different compositions of melter feed at different total solids loadings.

The above model was applied to simulated waste rheological data prior to frit addition, Figure 2. A reasonable match was obtained to rheological data following frit addition (11 volume percent frit) with f = 1.8.

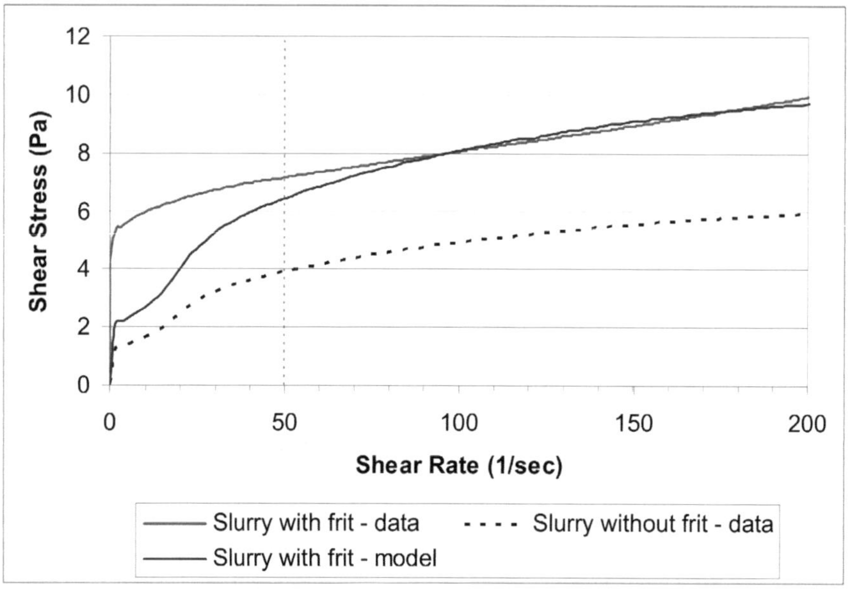

Figure 2. Application of the modified extended Einstein equation corrects rheological data for frit-free slurry to match that for waste slurry with frit at 11 volume percent.

The fit between the model and the slurry-frit data was better at higher shear rates (above 50/s). The shape of the raw flow curve data for the frit-free slurry had some unusual features below shear rates of 50/s. These irregularities were translated into the predicted melter feed flow curve.

Rheological modeling at SRS typically focuses on the region above 50/s shear rates, since most data are fit to the Bingham plastic equation over a range of several hundred reciprocal seconds of shear rate that are typical of site processes. An on-going program is currently evaluating the impact of replacing the glass frit with comparably sized glass beads, Figure 3.

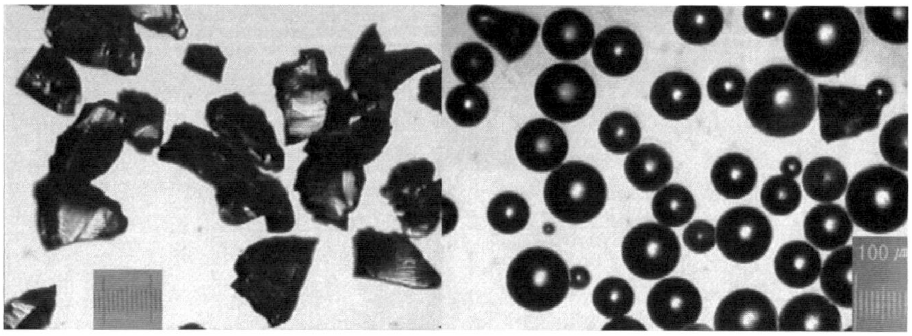

Figure 3. Comparison of glass frit (left) with chemically identical glass beads (right).

Improved rheological properties in the slurries with beads may permit the melter feed to be successfully processed with reduced water content (higher waste content, and less melter energy used for water evaporation).

A test of the modified Einstein equation was possible using flow curves for melter feed slurries with frit and with beads. Data were available for the same volume fractions of frit or beads in the same wt% total solids of HLW particles. The issue was whether or not the two flow curves could be reduced to a reasonably identical baseline flow curve describing the slurry without either particle. In this case an f value of 2.0 was needed for the flow curve with frit to match the result from the flow curve with beads using f equal to 1.0, Figure 4.

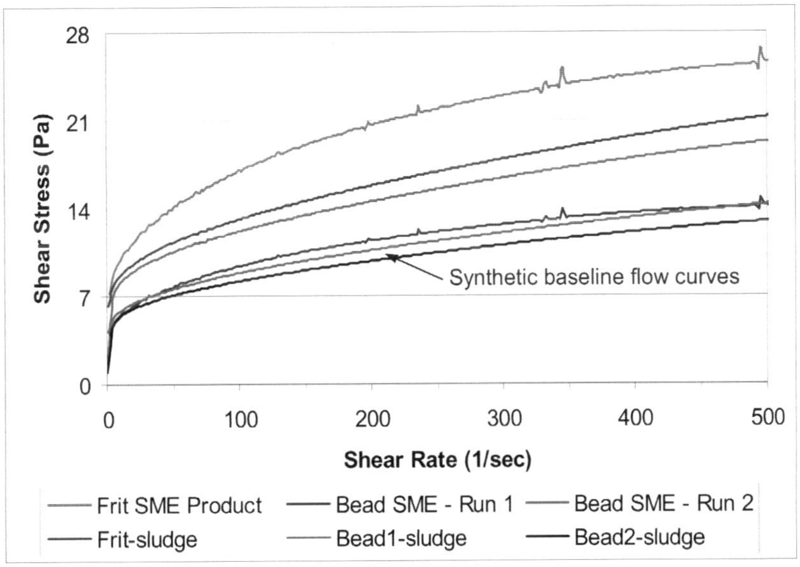

Figure 4. Reduction of the frit-based slurry flow curve data (red) and bead-based slurry data (blue) to a common baseline curve (green-brown-black).

The frit-based slurry was run in duplicate with negligible scatter, so only a single trace is shown in Figure 4. The variation in the bead replicate flow curve data is not unusual and is usually attributed to small variations in the insoluble solids content aliquoted from the sample vessel into the rheometer measuring cup. The spread in the three predicted baseline flow curves derived from the two different slurry systems is of similar magnitude to the general reproducibility of frit-slurry flow curves.

An emerging area of study is an effort to quantify and model the retention of bubbles in the settled solids in the waste tanks. The bubbles are H_2-O_2 mixtures from radiolysis of water. These gas molecules coalesce into bubbles within the settled liquid-particulate mass in the bottom region of the storage tanks. This represents another situation where heterogeneous, or multi-phase, models are likely to be more appropriate then homogeneous models. Molecular diffusion is not able to keep up with the radiolytic generation rate. The settled solid matrix appears to play a key role in retaining the bubbles. Rapid release of trapped bubbles occurs once slurry pumps are inserted and started in a tank containing settled solids. These bubbles can briefly lead to the formation of a flammable or explosive atmosphere over the tank contents, and that is unacceptable for nuclear waste storage.

It is believed that the retention of bubbles in settled HLW solids is not solely due to the more viscous properties that would prevail in the settled slurry due to the higher volume fraction of solids content than when the tank is fully mixed. The particles themselves are thought to try to orient at the bubble-slurry interface in such a manner that the bubbles become anchored to the settled matrix. Fundamental studies are needed to better understand the forces acting to prevent bubble coalescence and release.

REFERENCES

Koopman, D., D. Lambert, and M. Stone, "Successes and Emerging Issues in Simulating the Processing Behavior of Nuclear Waste Blends", *Proceedings of the 2009 AIChE Annual Meeting*, New York, NY: AIChE. 2009.

Guth, E., and R. Simha, *Kolloid-Z.* **74**, 266 (1936).

Thomas, D., *J. Colloid Sci.*, **20**, 267 (1965).

Govier, G. and K. Aziz, The Flow of Complex Mixtures in Pipes, p. 98, Van Nostrand Reinhold Co., New York, NY, 1972.

Solids Suspension by Hydrofoil Impellers Operating Close to the Vessel Bottom at Low Liquid Levels

Bob Dowd, CA, Philadelphia Mixing Solutions LTD, Palmyra, PA

Chris Hibshman, CA, Philadelphia Mixing Solutions LTD, Palmyra, PA

Wojtek Wyczalkowski, R&D, Philadelphia Mixing Solutions LTD, Palmyra, PA

Typically top entry agitators with turbine type high efficiency axial flow impellers are used to suspend and re-suspend slurry or blend fluids to continuously feed a process. The turbine impeller used by most agitator suppliers is an axial flow device pumping down to sweep the bottom of the tank and push solids up the side to maintain the suspension and blend well for a continuous process. While the axial flow impeller is currently the most efficient device to accomplish the task, the impeller requires a minimum distance off the tank bottom in order to perform, plus a minimum amount of fluid coverage above the impeller to establish impeller pumping. The resultant minimum fluid height in the tank can be as much as 30 percent of the tank diameter, potentially rendering 30% of the tank volume unusable, in a batch process, because the slurry suspension or blending cannot be maintained for pumping. To compound the issue, if the level gets too close or at the impeller, excessive hydraulic load will develop leading to mechanical failures. This presentation will present the problem and solution with calculations and video taped laboratory demonstrations.

See more of this Session: Mixing in Multi-Phase Systems - Solid-Liquid Systems
See more of this Group/Topical: North American Mixing Forum

Solids Suspension of a Mixture of Solids with Two Different Densities in Mechanically Agitated Tanks - Part 1. Low Concentration Effects

Most correlations for stirred tanks relate the speed to suspend solids to the solids concentration via a power law relation. Correlations like the Zwietering suggest that there is no minimum speed required to suspend solids, but many propose that there is a minimum speed below which solids can be suspended no matter how low the concentration.

A laboratory scale reaction vessel was used in this study with a Lightnin A310 impeller. A Mettler-Toledo Lasentec probe was then used to determine the solids distribution in a static stirred tank throughout the height of the vessel. Two solids with different physical properties were investigated: sand (density of 2656 kg/m^3) and urea formaldehyde (density of 1323 kg/m^3), at concentrations from 0.1 to 10.0 wt% in water.

Impeller speeds were investigated on a range of 900-1500 RPM for sand. It was established that using a higher agitation speed (1200-1500 RPM) increases the uniformity of the slurry concentration. It was also verified that the high concentrated region is independent of the liquid height in the tank.

Results show that the typical power law correlation where the exponent on the concentration is about the 0.15 power applies for the experimental setup studied in this case. Also, a comparison between weight fraction and volume fraction was made because theory suggests volume fraction has more mechanistic justification, but no significant differences were found. Finally, the data obtained for weight percentages below 1% were fitted to a linear equation that suggest a minimum speed below which particles cannot be suspended. With data between 0.1 and 1.0 weight percent it was not possible to distinguish which was a better fit.

Importance of Particle Properties On the Dispersion of Nanoparticle Clusters

Gustavo A. Padron, Fluid Engineering Centre, BHR Group Limited, Bedfordshire MK43 0AJ, United Kingdom

Gul Nerime Ozcan-Taskin, Fluid Engineering Centre, BHR Group Limited, Bedfordshire MK43 0AJ, United Kingdom

The dispersion of four types of nanoparticle clusters has been studied to investigate the kinetics and mechanisms of break up with. Results on the mechanisms and kinetics of break up obtained under different processing conditions will be reported.

See more of this Session: Mixing in Multi-Phase Systems - Solid-Liquid Systems
See more of this Group/Topical: North American Mixing Forum

Particle Transport in Closed Streamline Flows at Small Reynolds Number

Jeffrey F. Morris, Benjamin Levich Institute and Department of Chemical Engineering, The City College of New York, New York, NY

The transport of particles in closed streamline flows is examined through experiments and modeling of the bulk suspension flow. The experiments are performed primarily in a microfluidic environment at small conduit scale Reynolds number. We will report the averaged velocity determined from experiment and compare this with the predictions of the an established suspension flow model. The model is applied to predict the particle transport and compared to experimental results obtained from closed streamline flows formed by flow past an obstacle. The flow behavior at the extensional flow at the leading edge of the obstacle is also studied.

See more of this Session: Mixing in Multi-Phase Systems - Solid-Liquid Systems
See more of this Group/Topical: North American Mixing Forum

Temporally and Spatially Resolved Calculations of a Reacting Coal Jet Using Large Eddy Simulations (LES) and Direct Quadrature Method of Moments (DQMOM)

Jeremy Thornock, Philip Smith, Charles Reid, Julien Pedel

University of Utah, Institute for Clean and Secure Energy

Abstract

In a turbulent combusting coal jet, accurate prediction of the particle and velocity distributions are critical for predicting flame characteristics and ultimately heat flux in a coal-fired boiler. The present work combines a state-of-the-art particle model and turbulent-reacting flow model to provide a realistic simulation of the near burner region of a coal injector. The large eddy simulations (LES) technique is used to represent the turbulent coal jet the direct quadrature method of moments (DQMOM) is used to represent the particle phase. DQMOM provides an Eulerian particle-tracking method with full statistics. When, combined with the spatial and temporal resolution provided by LES, DQMOM yields a comprehensive statistical description of the behavior of the particle phase in the jet. Results are presented for a coal devolatilization experiment. Three key particle behaviors are note; particle segregation, particle clustering, and particle devolatilization. The present work identifies issues relevant to simulating reacting coal jets using LES and DQMOM, and lays the groundwork for future development.

1 Introduction

A major challenge in turbulence modeling is representing the large range of length and time scales. In general, modeling approaches separate these scales into two classes; resolved scales that are represented directly on the mesh and unresolved, or sub-grid scales that require modeling approximations. The Reynolds-averaging (RANS) approach, a familiar turbulence model in most industrial CFD packages, only represents the time averaged flow and as a result, models all turbulent scales. On the other hand, direct numerical simulations (DNS) seek to resolve all flow scales, yielding high fidelity results with little or no approximations. DNS is prohibitively expensive, however, for complex flows or moderate to large Reynolds numbers. Large eddy simulations (LES) present a middle ground; the large-scale flow structures are resolved without models, and only the small-scale turbulent flow structures are modeled. For flows with important time-dependent behavior, LES provides an accurate prediction of reality by resolving the large, unsteady scales, but vastly reduces the expense of DNS by modeling the small scales.

In our LES calculation, the coal particles are represented using an Eulerian method, which has a stationary frame of reference relative to the computational mesh. The goal of the current Eulerian method is to represent the spatial and temporal dependence [7]of the number density function

(NDF), which is the full statistical description of the particle phase. In particular, the direct quadrature method of moments (DQMOM) is used to represent the coal phase [3, 2, 5]. In the DQMOM approach, the NDF is represented using a quadrature approximation that itself is a set of delta functions. Computationally speaking, the DQMOM approximation results in a set of scalar equations representing the weights and weighted abscissa values of the quadrature approximation. These additional scalars are transported in an Eulerian manner on the computational mesh. Thus, the NDF of the particle phase is represented at little additional expense to the LES calculation. The current approach is made general enough that particle models that have been traditionally used in the Lagrangian approach are easily added to the DQMOM approach.

This work is focused on representing three important characteristics that occur in coal-combustion systems; 1) particle size segregation, 2) particle clustering and 3) particle devolatilization. All of these phenomena influence operational considerations within a coal-fired boiler such as flame stability, flame lift-off, and heat flux characteristic within the boiler. Of course, none of these effect in this multiphase system occur independently. They are a result of multi-physics interactions between phases and between mixing streams. These effects are inherently unsteady processes and the mean characteristics of the flow are a function of the time dependant behavior from which the mean was derived. Thus it behooves a successful model to have some representation of the unsteady effects present in the model itself, whether directly resolved or represented with a sub-model. For this reason, LES combined with DQMOM models are used in this work as they capture the unsteady phenomenon. The combination of the two provides spatial and temporal representations of the near burner region, including the particle phase statistics.

2 Methodology

2.1 LES Approach

The ARCHES [8]computational fluid dynamics tool is used here to simulate the reacting coal jet. The ARCHES code solves the LES filtered equations of mass, species, moment, and energy balances (with radiation) on a structured, finite volume mesh. Additionally, ARCHES uses parameter reduction techniques for reaction modeling coupled with sub-grid scale mixing models. The ARCHES code is a component of the Uintah Computational Framework (UCF)[6], a component-based problem solving environment that provides a general framework for massively-parallel simulations. The UCF manages the message passing and data handling needed for large-scale, parallel simulations performed on multi-processor computers. The simulations discussed in this work were run in parallel on processor numbers ranging from 324 to 832 which contained a set of grid cells on the order of 6M to 13M finite volumes.

2.2 Particle Representation

2.2.1 Number Density Transport Equation

The number density function (NDF) describes the number of particles per volume as a function of several particle independent variables (e.g., particle diameter, particle composition, etc.) called internal coordinates. These internal coordinates are contained in the internal coordinate vector $\boldsymbol{\xi}$ (boldface denotes vector quantities), and the internal coordinates are dependent on space and time $(\mathbf{x}, t)$.

The NDF transport equation is derived in the manner of Ramkrishna [7]. The velocity of the NDF in both real and internal coordinate space can be described using a set of internal coordinates $\boldsymbol{\xi} = [\xi_1, \xi_2, \cdots]$ and external coordinates $\mathbf{x} = [x_1, x_2, \cdots]$. The resulting transport equation of the NDF is

$$\frac{\partial f}{\partial t}+\sum_{i=1}^{3}\frac{\partial}{\partial x_i}\left(\langle u_{i,f}\rangle f\right)-\sum_{i=1}^{3}\frac{\partial}{\partial x_i}\left(D_{x_i}\frac{\partial f}{\partial x_i}\right)=-\sum_{j=1}^{N_\xi}\frac{\partial}{\partial \xi_j}\left(\langle G_{j,f}\rangle f\right)+\sum_{j=1}^{N_\xi}\frac{\partial}{\partial \xi_j}\left(\Gamma_{\xi_j}\frac{\partial f}{\partial \xi_j}\right)+h\left(\boldsymbol{\xi};\mathbf{x},t\right), \tag{1}$$

which represents the spatial and temporal evolution of the the NDF, including all particle phenomena (devolatilization, particle size change, breakage, etc.)

2.2.2 Quadrature Approximation: Direct Quadrature Method of Moments

A general N-point quadrature formula can be written as:

$$\int_a^b w\left(\mathbf{r}\right) g\left(\mathbf{r}\right) d\mathbf{r} \approx \sum_{\alpha=1}^{N}\left[w_\alpha\left(g\left(r_{1\alpha}\right) g\left(r_{2\alpha}\right)\ldots\right)\right] \tag{2}$$

The weights are common to all elements of $\mathbf{r}$ because the weight function $w\left(\mathbf{r}\right)$ is a single function common to all elements of $\mathbf{r}$.

The current NDF of interest (an NDF representing the coal particles) is multivariate and as such the direct quadrature method of moments (DQMOM) is used to solve Equation 1. The quadrature approximation from Equation 2 is applied to a multivariate NDF to yield,

$$f(\boldsymbol{\xi};\mathbf{x},t) \approx \sum_{\alpha=1}^{N}\left(w_\alpha(\mathbf{x},t)\prod_{j=1}^{N_\xi}\delta\left(\xi_j-\langle\xi_j(\mathbf{x},t)\rangle_\alpha\right)\right), \tag{3}$$

where N is the number of quadrature nodes and N_ξ is the number of internal coordinates. This is the quadrature-approximated number density function f. Next, if the moment transform of the quadrature-approximated NDF is taken, the integral disappears, and only an algebraic expression is left. Taking the moment transform of this multivariate NDF approximation yields the quadrature-approximated multivariate moment:

$$m_{\mathbf{k}} \approx \frac{\sum_{\alpha=1}^{N}\left\{w_\alpha\left(\prod_{j=1}^{N_\xi}\langle\xi_j\rangle_\alpha^{k_j}\right)\right\}}{\displaystyle\sum_{\alpha=1}^{N} w_\alpha} \tag{4}$$

where $\mathbf{k}$ is the multivariate moment index vector, and $\mathbf{k_j}$ is the jth element of $\mathbf{k}$ (with a total of N_ξ values, where N_ξ is the number of internal coordinates).

DQMOM solves transport equations for the weights and abscissas of the quadrature approximation. These transport equations are simply scalar transport equations, and are easily implemented into a CFD code. After the weights and weighted abscissas are obtained one can reconstruct moments of the NDF from Equation 4 to obtain statics on the particle phase such as particle size, coal mass fraction, particle number and so on. For details on the full DQMOM derivation, see [5].

3 Coal Models

DQMOM and LES are used to simulate a coaxial jet of particle-laden flow in the primary inlet and particle-free gas flow in the secondary inlet. Both inlet temperatures are 298 K and introduced into a

1800 K environment. The velocity ratio of the primary inlet to the secondary is 0.625. The diameter of the primary jet is 0.014m and the diameter of the secondary jet is 0.032. The simulation focuses on the near burner region.

The NDF in this work is a function of two internal (ξ) coordinates; particle size l, and particle mass of raw coal α_c. Each internal coordinate was tracked using two quadrature points in the quadrature approximation. The result is a set of six extra scalar equations for the CFD solver. The initialization for each weight (equivalent to particles for each environment per volume) was $w_0 = w_1 = 2\times10^{10}\,\text{vol}^{-1}$; the initialization for each internal coordinate was $l_0 = 35\mu\text{m}$, $l_1 = 75\mu\text{m}$ and $\alpha_{c0} = 0.91 \times 10^{-10}\,\text{kg}$, $\alpha_{c1} = 2.10 \times 10^{-10}\,\text{kg}$.

Coal devolatilization was modeled using the model of Kobayashi et al. [4]that describes the pyrolysis of coal as a function of temperature in the early stages of the combustor. In this paper, only the mass of raw coal α_c was tracked. Because of the inert environment, representation of the amount of char was not required. The conversion of raw coal to char, however, is represented in this devolatilization model.

Recently, Balachandar has applied the equilibrium Eulerian approximation to estimate the relative velocity between particles and the gas phase [1]. The relative velocity between the particles and the the gas is assumed to depend only on the particle's response time to the acceleration of the local fluid. If the particles are also assumed to be small enough that their response times are less than the Kolmogorov time scale (the smallest time scale of the fluid), then an expression for the relative velocity is obtain. See [1] for more details.

4 Results: Pyrolysis of Coal in the Near Burner Region

A snapshot of the mixture fraction (*kg hot gas/kg gas mixture*) is shown in Figure 1. The total domain length shown is roughly 55 primary diameters long. As expected, the mixture fraction goes to a fully mixed state as it moves down the reactor as the hot and cold gases mix. As will be shown below, the particle behavior only partially matches the gas-phase behavior and deviations are explained by the distribution of the local drag effects on the different particles sizes.

Results for a longitudinal slice across the jet radius, 20 primary diameters from the inlet, is shown in Figure 2. The plot shows the instantaneous number density of each delta function approximating the NDF per the quadrature approximation. We refer to the two delta functions approximating NDF as "environments" of the NDF. Because we have a quadrature of two, there are two environments (representing a small class of particles and a large class of particles). The results are collected from the LES simulation for a number of time steps. The black dots represent the smaller particles (Environment 0), which started at a particle size of $35\mu\text{m}$ and a raw coal mass of $0.19 \times 10^{-10}\,\text{kg}$, while the red crosses are the larger class of particles (Environment 1), which started at a particle size of $75\mu\text{m}$ and a raw coal mass of $2.10\times10^{-10}\,\text{kg}$. These values are the weights ($\#/vol$) of the two environments. Clearly, the smaller particles (black) have a higher number density near the edges of the jet, while the the larger particle (red) has a higher number density near the core of the jet. These results match the expected size segregation behavior of large and small particles in a turbulent jet. Additionally, particle clustering is observed for the Environment 0 across the radius of the jet as shown by the intermittent high concentrations of Environment 0 across the radius of the jet. Volume rendered images of the two particle fields are shown in Figures 3 and 4. These images reinforce the conclusion that the two particle classes are segregating and showing clustering effects.

Analogous results are shown for instantaneous values of mass of raw coal per particle in Figure 5. While the Environment 1 (red) will naturally have a larger amount of raw coal due to the fact that the particles simply have a larger mass, we observe that the value of α_c for the Environment 0 (black) reaches zero in a radial location in the vicinity of $0.05\,m$. However, there are still a substantial

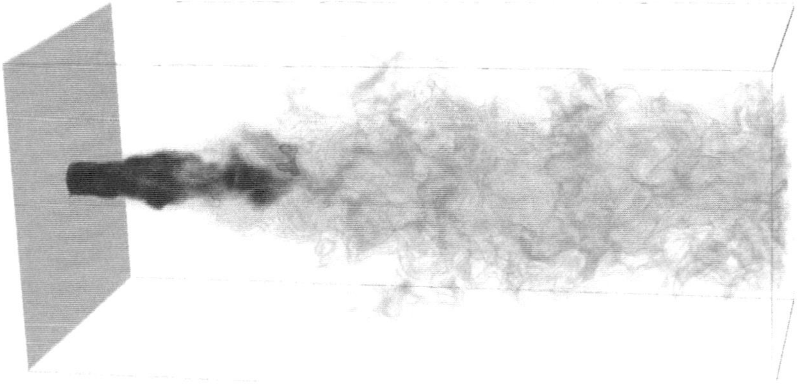

Fig. 1: Volume-rendered image of the gas mixture fraction (kg hot gas / kg mixture) throughout the reactor.

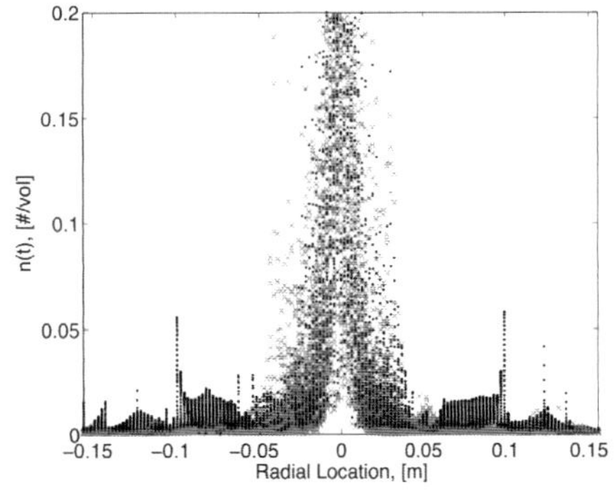

Fig. 2: Instantaneous number density results for coaxial jet at 20 diameters from the inlet. Results shown are number density for Environment 0 (black, smaller particles) and Environment 1 (red, larger particles).

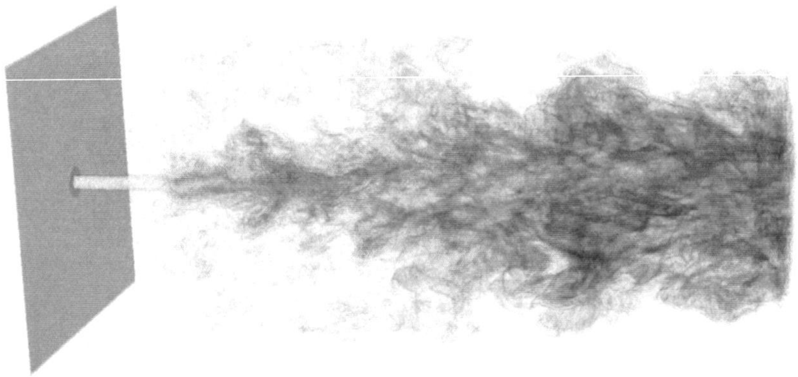

Fig. 3: Volume-rendered image of the number density of the smaller particles. The dispersion of the small particles closely resembles the dispersion of the gas mixture fraction as the smaller particles with little inertia tend to follow the gas streamlines.

number of smaller particles beyond this radius in the jet (see Figure 2). The small particles beyond the edge of the jet have completely devolatilized before the larger particles have uniformly devolatilized a significant amount. In other words, small particles move from the core of the jet to the edges, and begin to devolatilize first; the larger particles take longer to disperse, and devolatilize more slowly.

Using the environments of the NDF from the calculation, we construct the time averaged statistics of the coal-particle field NDF. This includes the mean value of the coal mass for the particle distribution shown in Figure 6. The coal particles initially devolatilize slowly, but as they increase in temperature from the entrainment of hot gas, the reaction rate accelerates. After much of the raw coal has reacted and disappeared, the reaction rate slows down again. A plot of the standard deviation of length down the centerline of the reactor is shown in Figure 7. This plot demonstrates a reduction of the standard deviation until roughly 15 diameters down stream. The reduction of the standard deviation of length is attributed to the particle size segregation. Little particles disperse quickly from the center of the jet leaving behind the larger particles with high interia, effectively narrowing the distribution of the NDF. Between 15 and 20 diameters the standard deviation remains fairly constant after which is begins to rise indicating a destruction of the potential core. Afterwards, the distribution widens again as the system becomes fully mixed, widening the NDF.

5 Conclusions

A formulation was presented for tracking the moments of the NDF of coal particles as the particles devolatilize in a hot jet using DQMOM. This formulation utilizes the quadrature approximation to approximate the NDF. The model formulation was kept general, so that existing Lagrangian models for the behavior of coal particles can be implemented for future DQMOM simulations.

Results were presented for a simulation of a devolatilizing coal jet using the direct quadrature method of moments. These results are very encouraging as DQMOM combined with the LES model produced clustering of larger particles, size segregation of different-sized particles, and differential

Fig. 4: Volume-rendered image of the number density of the larger particles. The jet is much less disperse compared with the number density of the smaller particles and compared with the mixture fraction of the primary inlet. Particle clustering is observed for this class of particles.

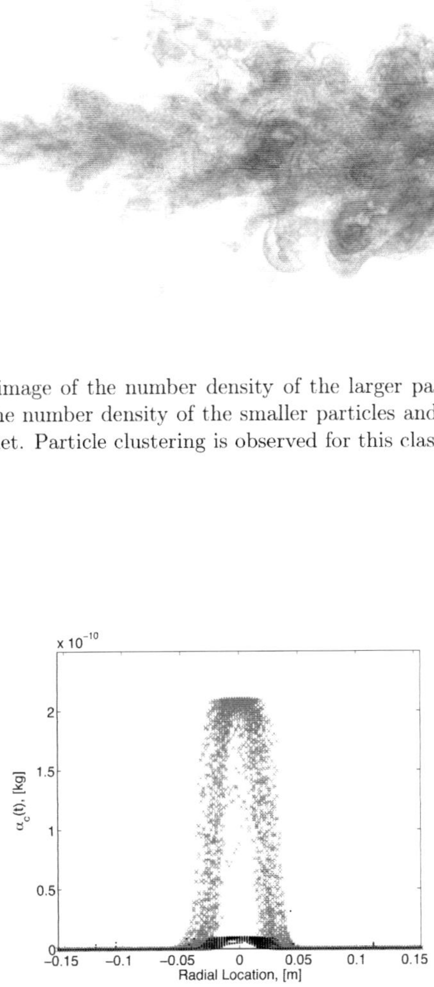

Fig. 5: Instantaneous values for α_c, the total mass of raw coal in each particle. Results are shown for the small particle (black) and the large particles (red).

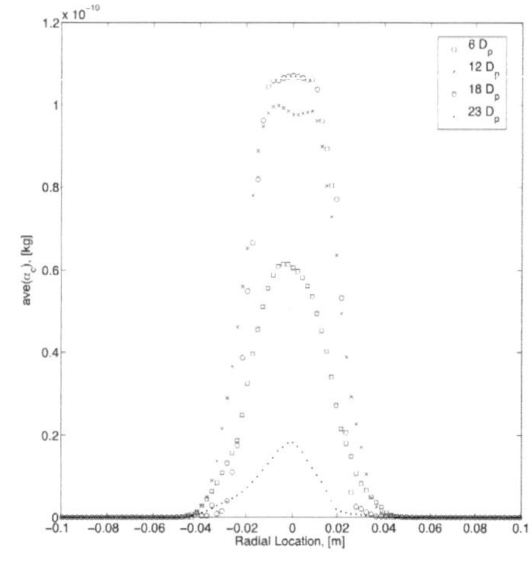

(a)

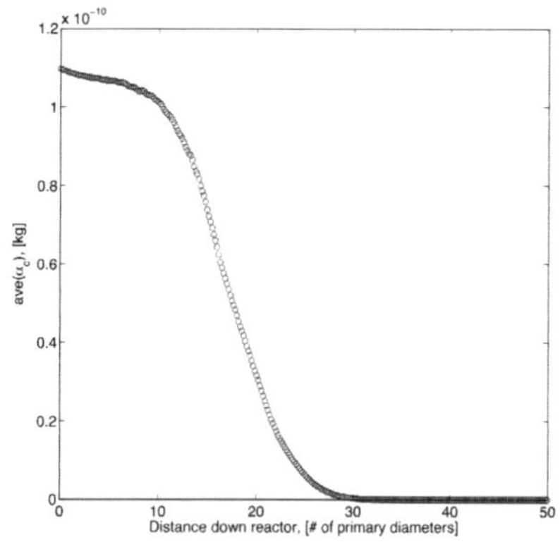

(b)

Fig. 6: Plots of the mean raw coal mass for the NDF showing a) the mass across the radius at different lengths down the reactor and b) the mass down the centerline of the reactor.

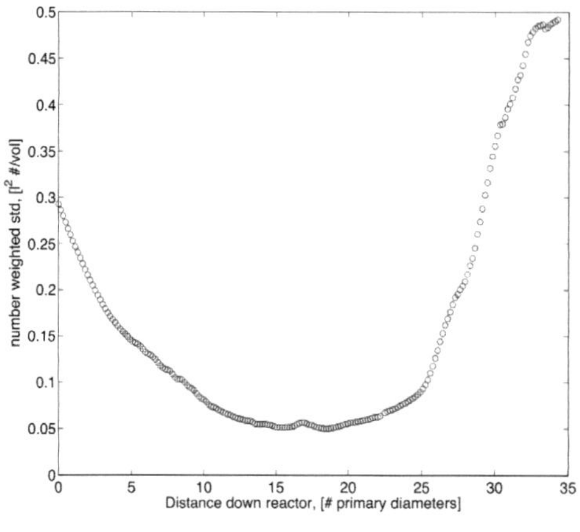

Fig. 7: The number weighted standard deviation of the particle length down the centerline of the reactor.

devolatilization behavior.

Future work includes introducing additional internal coordinates and adding additional particle models such as char oxidation, particle size changes, breakage, etc. to include more particle phsyics. Additionally, validation excersizes are need to test he model against experimental data. Finally, we desire to apply this model to oxy-coal systems to help assess flame stability issues in the near burner region for retrofit and scale-up purposes.

References

[1] S. Balachandar. A scaling analysis for point-particle approaches to turbulent multiphase flows. *Personal Communication*, December 2008.

[2] Rong Fan, Daniele L. Marchisio, and Rodney Fox. Application of the direct quadrature method of moments to polydisperse gas - solid fluidized beds. *Powder Technology*, 139:7–20, 2004.

[3] Rodney O. Fox. *Computational Models for Turbulent Reacting Flows.* Cambridge University Press, 2003.

[4] H. Kobayashi, J.B. Howard, and A.F. Sarofim. *Sixteenth Symposium (International) on Combustion*, pages 411–425.

[5] Daniele Marchisio and Rodney O. Fox. Solution of population balance equations using the direct quadrature method of moments. *Journal of Aerosol Science*, 36:43–73, 2005.

[6] S. Parker. A component-based architecture for parallel multi-physics pde simulation source. *Future Generation Computer Systems*, 22(1):204–216, 2006.

[7] Doraiswami Ramkrishna. *Population Balances.* Academic Press (San Diego), 2000.

[8] J.P. Spinti, J.N. Thornock, E.G. Eddings, P.J. Smith, and A.F. Sarofim. Heat transfer to objects in pool fires. In *Transport Phenomena in Fires.* WIT Press, 2007.

Mixing and Shear-Induced Migration in a 2D Time-Periodic Cavity-Driven Flow

Bu Xu, Lehigh University, Bethlehem, PA

James F. Gilchrist, Lehigh University, Bethlehem, PA

This study investigates mixing and segregation resulting from the competition between shear-migration and chaotic advection. In colloidal suspensions, particles do not behave as passive tracers. Local fluid deformation can result in cross-streamline migration due to the multi-body hydrodynamic interactions of particles, leading to concentration inhomogeneity. This behavior is studied numerically in a rectangular cavity flow with alternating periodic driving forces at top and bottom walls, a prototypical system for studying 2D chaotic advection (Ottino 1989). The combination of chaotic advection, repeatedly stretching and folding fluid elements, and diffusion enhance dispersion. Mixing and segregation, measured by the intensity of segregation, as a result of this interplay between shear-demixing and chaotic advection is highly complex. The effects from factors such as average volume fraction, particle radius, driving force and period are explored and compared to the underlying flow topology, as characterized by Poincaré maps.

See more of this Session: Research On Multi-Scale CFD for Particle Mixing and Suspension
See more of this Group/Topical: North American Mixing Forum

Particle Residence Time Distribution in a Continuous Flow Stirred Reactor: Experimental Characterization and CFD Modeling

Madan Somasi, The Dow Chemical Company, Midland, MI

Avinash Khopkar, Fluid Mechanics and Mixing, The Dow Chemical Company, Pune, India

Kishore Kar, Fluid Mechanics and Mixing, The Dow Chemical Company, Midland, MI

Nilesh Parmar, Fluid Mechanics and Mixing, The Dow Chemical Company, Pune, India

Ed Madenjian, Polyolefins Research, The Dow Chemical Company, Freeport, TX

Residence time distribution of catalyst particles in a continuous flow stirred vessel plays an important role in determining the final particle size distribution in processes where product is formed on catalyst particles. It is therefore necessary to control the mixing conditions in the vessel to improve the final particle size distribution. In this work, experimental characterization and a computational fluid dynamics (CFD) model were used to identify the appropriate mixing environment in a continuous flow stirred vessel to improve the final particle size distribution. Particle residence time distribution was first measured in a stirred vessel using two different vessel internal designs. A CFD model was then developed to predict the particle residence time distribution. The predicted particle residence time distribution was found to show reasonably good agreement with experimental data. The validated computational model was then used to evaluate different internal designs. It was observed that the particle size distribution could be made more plug-flow like by decreasing the off-bottom clearance of the upper two impellers and replacing the rotating discs with stationary discs (horizontal baffles) extended to the vessel wall. The resulting residence time distribution was demonstrated to more favorable than the existing design.

See more of this Session: Research On Multi-Scale CFD for Particle Mixing and Suspension
See more of this Group/Topical: North American Mixing Forum

Numerical Simulation of Marangoni Convection with Lattice Boltzmann Method Based On the Field Mediators

Yumei Yong, Key Laboratory of Green Process and Engineering, Institute of Process Engineering, Chinese Academy of Sciences, Beijing, China

Chao Yang, Multiphase Reaction Engineering, Institute of Process Engineering, Chinese Academy of Sciences, Beijing, China

Yi Jiang, Corning S. A. S, Avon F-77210, France

The Marangoni convection contributes to the renewal of the interfacial surface and is capable of increasing the mass transfer rate, resulting in the intensification of separation processes such as distillation, extraction, absorption, and desorption. Liquid-liquid mass transfer is showed experimentally dominating in gas-liquid and gad-liquid-solid catalytic reactions, especially in microchannels. Marangoni effect has been extensively studied for the case of liquid-liquid interface [1] and it is a promising topic to make full use of the Marangoni effect to enhance the mass transfer of two-phase flow in microchannels.

Computer fluid dynamics (CFD) can avoid the difficulties in experimental measurements and theoretical solutions. Over the last decade, the lattice Boltzmann method (LBM) has become an established numerical approach in CFD, because of its capability to simulate flow in multiphase systems. In particular, phase segregation and interfacial dynamics, which are essential in multiphase flows and difficult to be simulated by traditional approaches, can be modeled in LBM by the incorporation of molecular interactions. The sharp interface between different immiscible phases can be automatically maintained without any artificial treatment.

Therefore, the goal of this investigation is to gain insight into the mass transfer behaviour at liquid-liquid interfaces in microchannel. To investigate the mass transfer behaviour close to the liquid-liquid interface in microchannel, LBM based on the field mediators has been performed. Santos et al. [2] extended the field mediator concept to the Boltzmann models of immiscible fluids and the obtained predictions were compared with simulation results well. For popular LB methods, the chromodynamic, pseudopotential and the free-energy models can lead to unphysical behavior, such as the spurious current around interfaces, thermodynamic inconsistencies and the lack of Galilean invariance[3]. To some degree, LBM based on the field mediators can avoid the unphysical behavior because the LB model has been tested on its ability in describing the dynamical behaviors of the interface and Galilean invariance.

In a microchannel the mass transfer of different phases relies on the molecular diffusion under laminar flow, thus extraordinarily high efficiency and low liquid velocity are the main characteristics of mass transfer in a microchannel [4]. During the mass transfer process, the temperature gradient and concentration gradient can cause interfacial tension gradient and density gradient, by which Marangoni convection is induced. The treatment of interface between two phases is the most important to simulate accurately the Marangoni effect in microchannel, so the two-phase LB model is capable to deal with interface between two fluids in microchannel.

Now some scientists devoted to simulating the mass transfer process with LBM[5,6] but the numerical simulation of Marangoni effect with LBM is little, and it is far less than that of Marangoni effect caused by interfacial tension by other numerical methods.

In this paper, the Marangoni effect induced by interfacial tension gradient in the liquid-liquid two-phase system was investigated with LBM based on the field mediators. We first added the convection-diffusion equation of mass transfer into the two-phase LB model based on the field mediators, which describe accurately the dynamical behavior of the interface and Galilean invariance. The two-phase LB model coupled the convection-diffusion equation can capture the mass transfer and flow behavior on the interface between two phases. The influence of various process parameters (diffusion coefficients, concentrations, velocities and geometry) on the mass transfer was also explored. The quantitative numerical results of Marangoni convection induced by the interfacial tension gradient in the mass process was carried out and discussed in terms of mixing between two fluids to provide the flow information by comparison with the published data. The paper explored the influence of Marangoni effect on the mass transfer in two-phase systems and is helpful to explain the mechanism of that Marangoni effect induced by mass transfer.

References

[1]. Z.-S. Mao, J. Y. Chen, Numerical simulation of the Marangoni effect on mass transfer to single slowly moving drops in the liquid-liquid system. Chem. Eng. Sci., 2004, 59 : 1815

[2]. L. O. E. Santos, P. C. Facin, P. C. Philippi, Lattice-Boltzmann model based on field mediators for immiscible fluids, Phys. Rev. E., 2003, 68: 056302

[3]. X. Y. He, S. Y. Chen, R. Y. Zhang, A lattice Boltzmann scheme for incompressible multiphase flow and its application in simulation of Rayleigh-Taylor instability, J. Comput. Phys., 1999, 152: 642

[4]. Z. Yu, H. Orin, L. S Fan, Experiment and lattice Boltzmann simulation of two-phase gas-liquid flows in microchannels, Chem. Eng. Sci., 2007, 62: 7172

[5]. Q. M. Chang, J. I. D. Alexander, Study of Marangoni-natural convection in a two-layer liquid system with density inversion using a lattice Boltzmann model, Physics of Fluids, 2007, 19: 102107

[6]. J. Chen, X. F. Peng, D. J. Lee, Diffusion coefficient of Brownian particle in rough micro-channel, J. Colloid Interface Sci., 2006, 296: 737

See more of this Session: Research On Multi-Scale CFD for Particle Mixing and Suspension
See more of this Group/Topical: North American Mixing Forum

CFD Modelling for Gas-Liquid Two Phases of Single and Dual Impellers System

Ruey Chi Hsu, Chemical and material engineering, Chang Gung University, Kwei-Shan Tao-Yuan, Taiwan

Chung Tung Wu, Chemical and material engineering, Chang Gung University, Kwei-Shan Tao-Yuan, Taiwan

Both of fluid dynamic and gas dispersion in gas-liquid two phases stirred mixing system were simulated using Computation Fluid Dynamics (CFD). The FLUENT package with realizable k-e model and volume of fluid (VOF) model were used to carry out this simulation.

The simulation results were compared with the experimental results of single and dual Rushton impeller system. In single impeller system, the change of cavity structure agreed qualitatively well with previous experimental results, and the relationship between Pg/Pu and flow number agreed quantitatively well with previous studies. In dual impeller system, the co-current and counter-current rotation of impellers were studied. The simulated results of different relationship between Pg/Pu and flow number for different rotational pattern were both of qualitatively and quantitatively agreed well with experimental results. Furthermore, the different aeration results can be explained from the fluid dynamics.

See more of this Session: Research On Multi-Scale CFD for Particle Mixing and Suspension
See more of this Group/Topical: North American Mixing Forum

Aerobic Fermenter Scale-up: Why Equal Mass Transfer Scale-up Doesn't Always Work

by

Julian Fasano (Mixer Engineering Co.)
Eric Janz (Chemineer, Inc.)
Kevin Myers (University of Dayton)

AIChE Annual Meeting
November 8 – 13, 2009
Nashville, TN,
Paper 317a

In aerobic fermentation, scale-up has been traditionally thought to be mass transfer limited. We all know, but tend to forget, that there are other steps in the fermentation process regardless of the ultimate product. In addition to the transfer of oxygen from the gas phase to the liquid phase, the oxygen, along with nutrients, must be thoroughly mixed throughout the volume. Both oxygen and nutrients must diffuse through the cell wall, and then the cellular respiration takes place. This four-step rate process can be schematically viewed below:

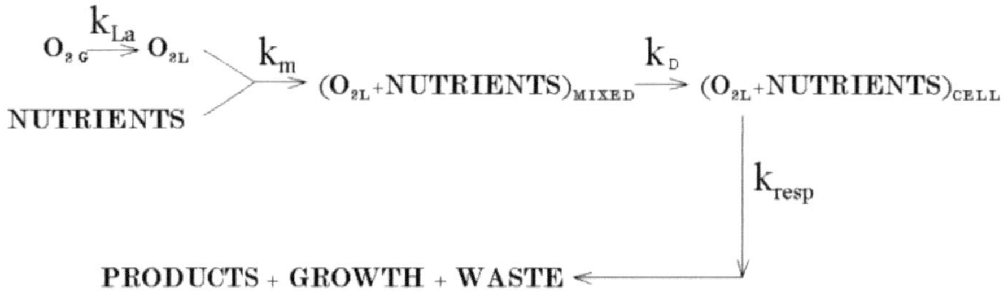

The desired respiration product of an organism can be highly dependent on a number of variables such as dissolved oxygen (DO), nutrient concentration, pH, etc. When these variables exceed certain bounds, even on a local basis, the behavior of the organism can be negatively affected in a number of ways.

Organism Sensitivity to Environment

One of the things that can go wrong is that organisms can die if environmental variable fall outside of a desirable range. Doble states "If at any point the cells have depleted their critical substrates or are starved of oxygen, an irreversible apoptosis may be triggered. The mixing time . . . has to be shorter than the time the cells take to deplete oxygen to the critical value . . ."[1] Schaechter and Lederberg[2] not only include O_2 but also pH and nutrient concentration. They also suggest that the respiratory rate can slow as well as organisms die when certain variables are outside of a critical range.

Namdev et al.[3] studied the effects of glucose fluctuations on synchrony in fed-batch fermentation of Saccharomyces cerevisiae. The fluctuations in glucose supply due to poor mixing were simulated by an intermittent supply of nutrients with a pulse-on time of 5 s and a pulse-off time varying from 3 to 39 s according to Monte Carlo method. Such rapid fluctuations in glucose supply during a fed-batch fermentation immediately produced a chaotic response in CO_2 levels which indicated a complete disruption of the synchrony established previously in the chemostat. They even went so far as to suggest that because large fermenters are inherently not well mixed that this process could not be scaled to large-scale fermenters.

Patnaik[4] studied the response coefficient of a fed-batch bioreactor to dissolved oxygen perturbation in complementary cultures during PHB (poly-β-hydroxybutyrate) production. In large bioreactors, incomplete mixing and perturbations in the dissolved oxygen (DO) concentration, both of which affect the fermentation, had to be considered. Ninety percent response times for all variables investigated were measured in hours.

Agarry et al.[5], in describing the substrate inhibition kinetics of phenol degradation by a binary mixed culture of *Pseudomonas aeruginosa* and *Pseudomonas fluorescence* from steady state and wash-out data, reported that kinetic parameters estimated from non-steady state experiments are often markedly different from those for steady state conditions. They found that non-steady state kinetics tend to over predict steady state phenol concentration, sometimes by more than an order of magnitude.

Excursions of DO or nutrients can also cause a pathway change to a different product. The Handbook of Food Biotechnology[6] states "As with nutrient concentration the DOT (dissolved oxygen tension, mg O_2/liter) may affect the metabolic pathway."

Summarizing, biological organisms are often metabolically sensitive to the surrounding environmental conditions. And if these conditions are not within certain bounds, even for short periods of time, organisms can:

1. Begin an irreversible apoptosis
2. Decrease their metabolism with the rate of recovery often measured in hours.
3. Produce products other than those intended.

The bioengineer needs to understand the sensitivity of the organism to a non-uniform environment before scale-up is undertaken. Some organisms are fairly robust and can withstand some variations in environmental parameters, and others cannot. For those organisms that cannot tolerate excursions beyond certain bounds, successful scale-up hinges upon ensuring that the environment is as uniform as possible by ensuring that the fermenter is well mixed.

Mixing Time and Equal Mass Transfer Scale-up

Mixing time can be estimated from a number of different techniques. Fasano and Penney[7] provided a technique with impeller specific parameters. The 99% homogeneous blend time can be estimated from:

$$t_{99} = \frac{4.605}{aN\left(\frac{D}{T}\right)^{b}\left(\frac{T}{Z}\right)^{0.5}}$$

where *a* and *b* are impeller dependent constants. They provide constants for various impellers including a radial flow impeller, a mixed flow impeller and a high efficiency axial flow impeller. There is a correction factor when the impeller Reynolds falls below 10,000. With very few exceptions aerated fermenters are generally well into the turbulent range and no blend time correction is required. Grenville and Nienow[8] provided a more general technique based on impeller power number, N_p. Their blending equation for turbulent flow is:

$$t_{95} = \frac{5.20}{N N_p^{1/3}} \left(\frac{T}{D}\right)^2 \left(\frac{Z}{T}\right)^{0.5}$$

Their transitional regime blend time equation becomes a function of the Reynolds number.

$$t_{95} = \frac{183^2}{N N_p^{2/3} N_{\mathrm{Re}}} \left(\frac{T}{D}\right)^2 \left(\frac{Z}{T}\right)^{0.5}$$

In both approaches, in turbulent flow when the impeller style, the impeller speed, the impeller power number, the ratio to T/D and Z/T remain constant, the blend time is the same regardless of scale.

The relationship of agitator speed to scale for equal mass transfer scale-up can be developed by maintaining constant k_La as follows:

$$k_L a \propto \left(\frac{P}{V}\right)^{2/3} \propto \left(\frac{N^3 D^5}{T^3}\right)^{2/3} \propto \left(\frac{N^3 T^5}{T^3}\right)^{2/3} \propto \left(N^3 T^2\right)^{2/3} \propto N^2 T^{4/3}$$

When conducting an iso-geometric scale-up for equal mass transfer coefficient, the impeller speed will vary with the tank diameter as

$$N \propto T^{-2/3}$$

Therefore, when scaling-up via equal mass transfer, speed is reduced as scale increases. And when speed goes down, blend time goes up. The effects of an iso-geometric scale-up on blend time are shown in Table 1.

Table 1: Equal Mass Transfer Iso-Geometric Scale-Up, T/Z=3.9

Tank Diameter		Volume		Motor Power		RPM	$t_{99.9}$ Blend Time, sec.
inches	meters	gallons	m^3	horsepower	kW		
20	0.51	104	0.394	1	0.746	354	47
49	1.24	1540	5.84	15	11.2	193	85
116	2.95	20400	77.3	200	149	107	154
270	6.86	258,000	978	2500	1865	60	282

These blend times are estimated for an un-gassed five Rushton or D-6 impeller system and are intended to demonstrate what happens to blend time when equal mass transfer scale-up is utilized. In addition to the blend time parameters demonstrated above, blend time in multiple impeller systems is also a function of impeller spacing and pumping direction. Different impeller systems, even at the same input power and speed, can exhibit blend times that differ by a factor of three. And blend times under gassed conditions can vary significantly from un-gassed blend times.

Techniques for Assessing the Effect of Blend Time

Before an equal mass transfer scale-up is undertaken, the bio-engineer should assess the effect of blend time. As a review, the equation for mass transfer rate is

$$M = k_L a(C^* - C)$$

$$\begin{aligned} \text{where } M &= \text{rate of mass transfer, kg/(s} \bullet m^3) \\ k_L a &= \text{volumetric mass transfer coefficient, } 1/s \\ C^* &= \text{oxygen saturation concentration, } kg/m^3 \\ C &= \text{oxygen concentration, } kg/m^3 \end{aligned}$$

The mass transfer coefficient, $k_L a$, is normally correlated as a function of power per volume and superficial gas velocity using an equation of the following form

$$k_L a = k\left(\frac{P_g}{V_l}\right)^{a'} v_{sg}^{b'}$$

$$\begin{aligned} \text{where } k &= \text{experimentally determined constant} \\ P_g &= \text{gassed power, kW or Hp} \\ V_l &= \text{liquid volume, gal or } m^3 \\ v_{sg} &= \text{superficial gas velocity, ft/s or m/s} \end{aligned}$$

The proportionality constant k and the exponent constants a' and b' can be significantly influenced by liquid properties. Also, compounds such as surfactants and insoluble oils can affect these constants as well. This correlating equation for $k_L a$ is valid only when the impeller(s) is not flooded. These constants should be experimentally determined on a smaller scale with a similar chemical and bio-chemical environment prior to scale-up. Bakker et al.[9] suggest that for an air-water system the constants can be approximated as: $a' \cong 0.6$ and $b' \cong 0.6$ and $k \cong 0.015 \pm 0.005$.

If the biology will permit, the easiest way to assess the effect of blend time, is to slow down the mixer, and increase the partial pressure of oxygen such that M, the mass transfer rate, remains constant while blend time increases. Either increasing the pressure within the fermenter or increasing the concentration of oxygen in the feed or a combination of both can accomplish this. It is recommended that the speed be reduced at least 30% each time a new run is made. Of course this will only work as far as the dispersing impeller is not flooded.

Efficient Gas Dispersing and Mixing Systems

Single impeller systems are generally only used for laboratory studies and in very small production scales. Fermenters of any reasonable size will typically have large liquid height to tank diameter ratios (Z/T) that will require multi-impeller systems. There are some very large differences in blend time for the various mixing systems currently employed industrially.

The authors observed over 20 years ago, that only a single radial dispersing impeller was required for adequate gas dispersion. Adding additional radial dispersing impellers for tall batches did not significantly enhance mass transfer. However, the addition of axial flow impellers instead, prevented significant coalescence and enhanced blending.

Blending demonstrations were conducted using three different impeller systems. The first system is an all D-6 impeller system. This system was very popular in fermenters from the early 1950's to the late 1980's.

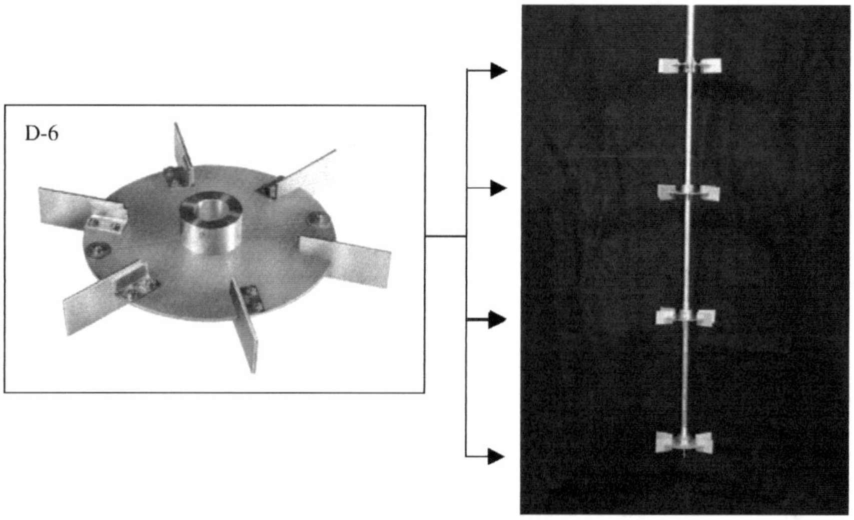

Figure 1: Rushton or D-6 Impeller System

Another impeller system that was demonstrated was the HE-3/CD-6 impeller system. This system was used predominately from the late 1980"s to the late 1990's. Additional information on this system is available in an article by Bakker and others.[10] It provided a dramatic improvement in blend time.

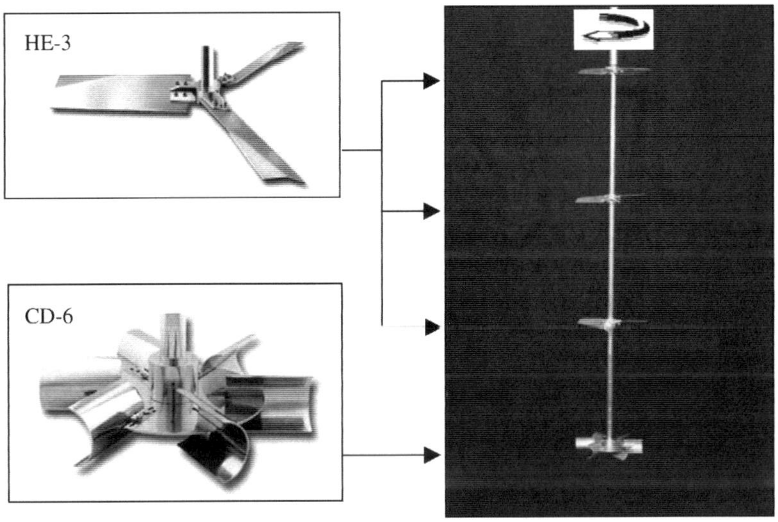

Figure 2: HE-3 and CD-6 Impeller System

The CD-6 (lowest impeller) adequately and efficiently dispersed the gas and the high efficiency HE-3's (upper impellers) efficiently pumped and reduced the mixing time of the fermenter contents.

The third impeller system demonstrated was the Maxflo W/BT-6 impeller system. The BT-6 impeller is a highly efficient gas dispersing impeller. The gassed power drop is only slightly more than the power drop expected from the effective un-gassed to gassed fluid specific gravity change. The Maxflo W impellers have a low tendency to cavitate, pump up and do not fight but enhance the natural tendency for the rising gas to carry fluid with it.

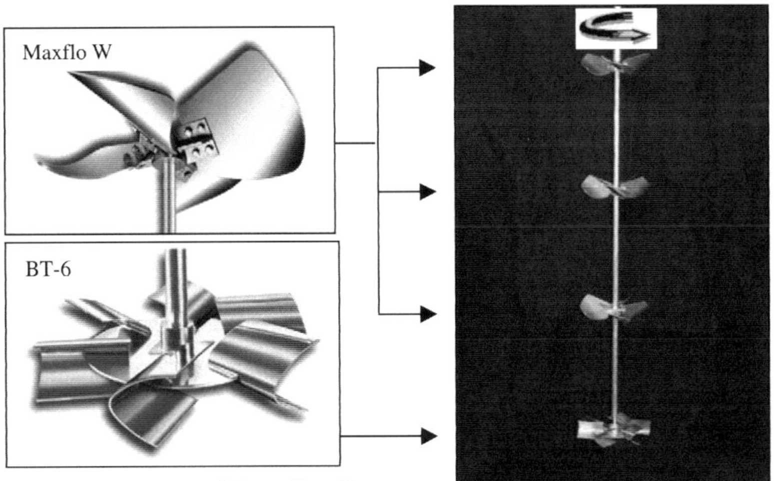

Figure 3: Maxflo W and BT-6 Impeller System

A general arrangement of the test system in shown below:

Figure 4: Blend Time Test System

Two hundred fifty ml of 0.05% thymolphthalein acid/base indicator was used. The color change was a blue (basic) to clear (acidic). The acid was added midway between the shaft and tank wall and midway between baffles on the surface. The time to color change was made both with an electric timer (pictured) and a stopwatch. The average time was used. Two runs were made for each case. The table below describes the various characteristics of the test system.

Table 2: Characteristics of Blend Time Demonstration System

Tank ID, in	15.5
Tank Straight Side, in	46
Dished Bottom Depth, in (approx.)	3.5
Liquid Level, in	43.4
Liquid Volume, gal	34.39
Liquid Level, in (prior to gas)	40.5
Gas Holdup, %	7.0%
Liquid Holdup Height, in	2.75
Indicator Concentration, %	0.20%
Base (NaOH) Concentration, N	5.0
Acid (HCl) Concentration, N	5.0
Base Added above Neutral, ml	30
Acid Added for Blend Time, ml	60

The technique for determining blend times was presented by Fenic and Fondy[11]. The time to achieve blending is referred to as the blend time, and adding twice the stoichiometric amount to achieve the color change produced a blend time of one half of a stoichiometric addition. For the most part, the blend time determination technique of Fenic and Fondy was also utilized in this demonstration.

Runs were made with the following impeller systems described below:

Table 3: Impeller Systems Demonstrated

Impeller Systems	Off-Bottom, in	Impeller Spacing, in.
(4) 5.7" D-6 Impellers (Figure 1)	4.86	10.5
(1) 5.7" CD-6, (3) 6.5" HE-3 Impellers (Figure 2)	4.86	10.5
(1) 6.5" BT-6, (3) 6.375" Maxflo W Impellers (Figure 3)	4.86	10.5

Since specific mass transfer rate is highly related to the power invested per unit volume, each run was made at equal power. The speed was increased until a power of 0.075 hp was reached. This is consistent with a power per unit volume of approximately 2 hp per 1000 gallons. The data for these runs is tabulated below:

Table 4: Blend Time Data

Run No.	Impeller System	Speed, rpm	Air, SCFM	Superficial Gas Vel. ft/sec	Partial Blend Time, sec.	Mean Partial Blend Time, sec	Stoichiometric Blend, Time sec.
1	(4) 5.7" D-6	203	0	0	29.8	28.7	57.4
2	(4) 5.7" D-6	203	0	0	27.6		
3	(1) 5.7" CD-6, (3) 6.5" HE-3	336	0	0	10.8	9.7	19.3
4	(1) 5.7" CD-6, (3) 6.5" HE-3	336	0	0	8.5		
5	(1) 6.5" BT-6, (3) 6.375" Maxflo W	270	0	0	10.0	9.6	19.1
6	(1) 6.5" BT-6, (3) 6.375" Maxflo W	270	0	0	9.1		
7	(4) 5.7" D-6	245	2.5	0.03	13.1	12.9	25.8
8	(4) 5.7" D-6	245	2.5	0.03	12.7		
9	(1) 5.7" CD-6, (3) 6.5" HE-3	375	2.5	0.03	6.3	6.7	13.3
10	(1) 5.7" CD-6, (3) 6.5" HE-3	375	2.5	0.03	7.0		
11	(1) 6.5" BT-6, (3) 6.375" Maxflo W	276	2.5	0.03	6.2	6.6	13.1
12	(1) 6.5" BT-6, (3) 6.375" Maxflo W	276	2.5	0.03	6.9		

Two observations can be made from the data:

1. The introduction of gas reduces blend time when holding the power constant. This was especially true where the gas helped minimize zoning in the Rushton or D-6 impeller system. This should not be an unexpected result. The total power to the fluid was roughly doubled when the gas was added. And with more invested power one would expect the blend time to be reduced. Middleton and Smith report that if agitator power is not held constant and gassed agitator power falls off too much, blend times may increase.[12]

2. Impeller systems can make a dramatic difference in blend time. The mixed impeller systems were significantly better than the all Rushton system due to their good top-to-bottom mixing that eliminates impeller staging.

By using a highly efficient blending and mass transfer impeller system, problems arising from blend time with scale-up can be forestalled to a larger scale.

<u>Summary and Conclusions</u>

For some biologies, when blending of dissolved oxygen, nutrients, or pH modifiers takes too long, organisms can:

1. Begin an irreversible apoptosis
2. Decrease their metabolism with the rate of recovery often measured in hours.
3. Produce products other than those intended.

The bio-engineer should conduct studies to determine whether blending will cause a problem during scale-up. Some biologies are robust and can tolerate longer blend times.

There are potentially some techniques that can be used to determine whether certain biologies will pose a problem on scale up because of increased blend time. The easiest method is to slow down the speed of the agitator but increase the partial pressure of oxygen so that mass transfer rate remains constant while blend time is increased.

There are mixed impeller systems, especially the Maxflo W – BT-6 system is increased, that are efficient in blending as well as mass transfer. Such systems can forestall the need for scale-up by equal blend time instead of equal power per unit volume.

When scaling-up fermenters there can be a limiting size above which equal power per unit volume scale-up will not work. This paper discussed a potential means of assessing this, and suggested impeller systems that could extend the limiting size.

Nomenclature

a = constant in mixing time equation or interfacial area per unit volume

a', b' = constants in $k_L a$ equation

C = dissolved oxygen concentration

C^* = *dissolved* oxygen saturation concentration

D = *impeller* diameter

k_D = diffusion rate coefficient

$k_L a$ = *volumetric mass* transfer coefficient

k_m = mixing rate coefficient

k_{resp} = *respiration* rate coefficient

M = specific mass transfer rate

N = impeller rotational speed

N_p = *impeller* power number = $P/\rho N^3 D^5$

N_{Re} = *impeller* Reynolds number = $\rho ND^2 / \mu$

P = impeller power

P_g = *impeller* gassed power

t_{95} = *time* to achieve 95% blend uniformity

t_{99} = time to achieve 99% blend uniformity

T = tank diameter

v_{sg} = superficial gas velocity

V_l = volume of liquid

Z = liquid height

μ = fluid viscosity

ρ = fluid density

References

1 Doble, Mukesh, "Avoid the Pitfalls of Bioprocess Development", CEP, August 2006, pp 34-41.

2 Schaechter, Moselio, Lederberg, Joshua, "The Desk Encyclopedia of Microbiology", 2nd Edition, Academic Press, 2004, p. 614.

3 Namdev, P. K., Thompson, B. G., Ward, D. B., Gray, M. R., "Effects Of Glucose Fluctuations On Synchrony In Fed-Batch Fermentation Of *Saccharomyces Cerevisiae"*, Biology Technology Progress, Volume 8, Issue 6, 5 September 2008, Pages 501 – 507.

4 Patnaik, P. R., "Response Coefficient Analysis of a Fed-Batch Bioreactor to Dissolved Oxygen Perturbation in Complementary Cultures During PHB Production", Journal of Biological Engineering, 27 March 2008.

5 Agarry S. E., Solomon, B. O., and Layokun, S. K., "Substrate Inhibition Kinetics Of Phenol Degradation By Binary Mixed Culture Of *Pseudomonas Aeruginosa* And *Pseudomonas Fluorescence* From Steady State And Wash- Out Data", African Journal of Biotechnology, Vol. 7 (21), pp. 3927-3933, 5 November, 2008, pp. 3927-3933.

6 NIIR Board of Consultants and Engineers, "Handbook of Food Biotechnology, Chapter: Fermenter and Bioreactor Design", National Institute of Industrial Research, 2005

7 Fasano, J. B., Penney, W. R., "Avoid Blending Mix-ups", Chemical Engineering Progress, October 1991, pp. 56-63.

8 Grenville, R. K., Nienow, A. W., "Handbook of Industrial Mixing: Science and Practice", Chapter 9: Blending of Miscible Liquids, Edited by Paul, E. L., Atiemo-Obeng, V. A., and Kresta, S. M., John Wiley and Sons, 2004.

9 Bakker, A., Smith, J.M., Myers, K.J., "How To Disperse Gases in Liquids", Chemical Engineering, Volume 101, Number 12, December 1994, pp. 98-104.

10 Bakker, A., Fasano, J. B., Myers, K. J., "Gas Dispersion Using Mixed High-Efficiency/Disc Impeller Systems", Presented at the 8th European Conference on Mixing, Queens College, Cambridge, England, 21-23 September 1994, published in IChemE Symposium Series 136, pp. 65-72.

11 Fenic, J.G., Fondy, P.L., "Application of Similarity Analysis to Blending of Miscible Liquids", Presented at the Annual AIChE Meeting , Atlantic City, NJ, Sept. 18-21, 1966, paper No. 4d.

12 Middleton, J. C., Smith, J. M., "Handbook of Industrial Mixing: Science and Practice", Chapter 11: "Gas-Liquid Mixing in Turbulent Systems", Edited by Paul, E. L., Atiemo-Obeng, V. A., and Kresta, S. M., John Wiley and Sons, 2004.

Mixing Time in a Stirred, Torispherical-Bottomed Tank Equipped with a Retreat-Blade Impeller Under Unbaffled and Partially Baffled Conditions

Nonjaros Chomcharn, Otto H. York Department of Chemical, Biological and Pharmaceutical Engineering, New Jersey Institute of Technology, Newark, NJ

Piero M. Armenante, Otto H. York Department of Chemical, Biological and Pharmaccutical Engineering, New Jersey Institute of Technology, Newark, NJ

Billy R. Allen, Lilly Corporate Center, Eli Lilly and Company, Indianapolis, IN

Glass-lined reactors are typically used in the pharmaceutical and specialty chemicals industries because of their material compatibility with most reactants and their cleanability. These reactors are typically equipped with a retreat-blade impeller placed close to the bottom of the tank a single baffle mounted from the top. The mixing performance of such reactors has not received significant attention in the literature, despite their ubiquitous presence in the pharmaceutical industry. In particular, mixing time, i.e., the time required by the system to achieve a predefined level of homogeneity, has not been studied to any significant degree.

In this work, the mixing time of a 60 L, scale-down version of an actual industrial reactor having an industrial-type torispherical bottom and equipped with a retreat-blade impeller was experimentally determined for both the unbaffled case and the case in which the typical beavertail baffle was used (i.e., under partial baffling). A conductivity method using sodium chloride (NaCl) as a tracer was used as one of the approaches to determine mixing time. Experiments under unbaffled system were conducted by installing one two probes in the mixing vessel (at the wall and midway between the wall and the shaft, respectively). The presence of the conductivity probe(s) had a significant impact on mixing time. Results showed that different mixing times were obtained with the conductivity method depending on the location of the probe(s) and the number of probes.

A separate colorimetric method coupled with image processing was additionally used to determine the mixing time. Both methods produced similar results when two probes were present, probably because of the baffling effects introduced by the probes themselves. Results were also obtained with the partial baffling arrangement, but only using the conductivity method. Experiments were performed when the agitation speed was varied in order to establish correlation between mixing time and rotational speed under both unbaffled and partially baffled systems. The mixing time in partially baffled system appears to be inversely proportional to the impeller speed and the predicted mixing time from regression agrees well with the experimental results. The non-dimensional mixing number, θ95N, in partially baffled system was obtained within the speed range of 100-200 rpm.

See more of this Session: Mixing in the Pharmaceutical Industry
See more of this Group/Topical: North American Mixing Forum

Computational Fluid Dynamics Simulation of a Quadrupole Magnetic Sorter Flow Channel

V. S. K. Sajja, Chemical Engineering, Auburn University, Auburn, AL

David J. Kennedy, Ikotech LLC, New Albany, IN

Paul W. Todd, Techshot, Inc., Greenville, IN

Thomas R. Hanley, Chemical Engineering, Auburn University, Auburn, AL

Isolation of the pancreatic islets is the important step in preparing islets for the transplantation in type 1 diabetes patients. Quadrupole magnetic flow sorting (QMS), effective in the separation of suspensions of single cells, is being modified for isolation of pancreatic islets. The separating mechanism in QMS is created by the magnetic field strength and gradient induced by a modified Halbach quadrupole magnet assembly and applied to an annular channel of separands flowing through the magnet bore. Thus, islets are subjected to vertical parallel annular flows, horizontal magnetic force, vertical sedimentation and flow through inlet and outlet spreaders and splitters as well as through peristaltic pumps. Computational fluid dynamics (CFD) models were used to simulate the islet separation in the flow channel and flow distributors in the QMS channel. Analyses of particle trajectories, splitter geometries and transport lamina were performed. The CFD results provided an adequate match with the experimental data.

Keywords: magnetic flow sorter, pancreatic islets, computational fluid dynamics, pulsatile flow

See more of this Session: Mixing in the Pharmaceutical Industry
See more of this Group/Topical: North American Mixing Forum

Development of a Mixing Model in the Scale-up of An API Crystallization

Adam M. Sokolnicki, Altus Pharmaceuticals, Waltham, MA

The conversion to product in a reactive-mixing system is governed in part by the scale and intensity of feed component segregation. During fed-batch anti-solvent crystallization, segregation drives the local-scale supersaturation, the magnitude of which influences nucleation and subsequent growth of crystals. Furthermore, the solvating ability of the anti-solvent system controls the nucleation order and dictates the required mixing intensity. This case study examines an approach to understanding and characterizing the role mixing plays during the development and scale-up of a drowning-out crystallization for a biopharmaceutical enzyme. Lasentec FBRM provided real-time feedback during ranging and scale-up studies, while a model competitive-parallel reaction system was adopted to gain insight into the feed-mixing process. These tools provided for a successful scale-up from the 20g bench scale to 30kg engineering runs.

See more of this Session: Mixing in the Pharmaceutical Industry
See more of this Group/Topical: North American Mixing Forum

Micromixing in Stirred Tanks Equipped with a High-Shear Homogenizer

Micaela Caramellino, Otto H. York Dept. of Chemical, Biological and Pharmaceutical Engineering, New Jersey Institute of Technology, Newark, NJ

Piero M. Armenante, Otto H. York Department of Chemical, Biological, and Pharmaceutical Engineering, New Jersey Institute of Technology, Newark, NJ

Stirred tanks equipped with both a centrally mounted primary impeller and an additional high-shear homogenizer are commonly encountered in industrial practice, and especially in the pharmaceutical industry. While the primary impeller generates bulk flow, the high-shear device is often added in an off-center position to generate a high-shear zone and enhance operations such as solid dispersion and liquid emulsification.

In this work, the effect of mixing in a stirred tank equipped with a Rushton turbine and a high-shear rotor-stator homogenizer was quantified using competitive reactions (third Bourne reaction set). The limited reactant (NaOH) was slowly added to the tank in which the other reactants were present from the beginning (HCl and ethyl chloroacetate) and where two parallel competing reactions took place. The yield of one the products (ethanol) was measured at different experimental conditions, and was used to quantify the effectiveness of micromixing.

The effect of a number of operating parameters was studied, including the speed of the primary impeller and the speed of the high-shear homogenizer. In addition, the effect of the injection point of the limited reactant was examined. For this purpose, a new injection device was designed to allow the introduction of the limiting reactant at a specific point near the homogenizer. The product yield was then compared with the case in which the injection was made at the liquid surface. Experiments were conducted in two tanks of different sizes to establish the effect of scale on the mixing process. In general, the results were significantly affected by different operating conditions. A significant decrease in mixing time, measured separately, was also observed when the new injection device was used, which confirmed the effectiveness of the homogenizer as a mixing intensifying tool when used in regular mixing tanks.

See more of this Session: Mixing in the Pharmaceutical Industry
See more of this Group/Topical: North American Mixing Forum

Influence of Mixing On the Dissolution Performance of a Pharmaceutical Dosage Form

Otute Akiti, Process Engineering, CMC, Exelixis, Inc., South San Francisco, CA

Sean P. Cusack, Process Engineering, CMC, Exelixis, Inc., South San Francisco, CA

Deana Tolentino, Formulation Development, CMC, Exelixis, Inc., South San Francisco, CA

Rhea Donley, Formulation Development, CMC, Exelixis, Inc., South San Francisco, CA

Dissolution testing is a characterization technique employed in the pharmaceutical industry to design formulations and assess product quality. It is a required performance test by many regulatory authorities for solid oral dosage forms, and is unique in that it is intended to measure the effect of the formulation and physical properties of the active pharmaceutical ingredient (API) on the in vitro rate of drug dissolution. In a typical pharmaceutical setting, the API is synthesizes and manufactured by one team and then handed over to subsequent team which formulates the API into a dosage form. Since the dissolution characteristics of the dosage form depend not only on how it is formulated, but also on the API's powder properties, a close collaboration between the two teams is necessary. In this contribution, we present the results of the collaborative efforts between the API process development team and the formulation development team to assess and produce a product with consistent dissolution characteristics which meet established performance specifications. We also demonstrate in this work the significance of mixing on achieving the established process goals.

See more of this Session: Mixing in the Pharmaceutical Industry
See more of this Group/Topical: North American Mixing Forum

Mixing Scale-Down Studies for the Scale up of Beer Fermentation

Alvin W. Nienow, Centre for Bioprocess Engineering, University of Birmingham, Birmingham, United Kingdom

Georgina McLeod, Centre for Bioprocess Engineering, University of Birmingham, Birmingham, United Kingdom

Christopher J. Hewitt, Centre for Biological Engineering, Chemical Engineering, Loughborough University, Loughborough, United Kingdom

Beer fermentations have traditionally been undertaken without the use of mechanical agitation with mixing being provided only by the CO2 evolved during the batch process. This approach has largely been maintained because of the belief in industry that rotating agitators would damage the yeast. In this work at the bench scale, we show that yeast is very robust and is able to withstand very intense agitation intensity under aerobic conditions without observable damage as measured by flow cytometry and other parameters. Much less intense agitation is also shown to reduce batch fermentation time for anaerobic beer production by about 25% compared to mixing by CO2 evolution. It is also shown that spatial variations in yeast concentration and temperature that exist when mixing by CO2 leads to slower fermentations and lower productivity. These bench scale studies have recently been supported by the work of others at the commercial scale from mixing by an impeller or by a rotary jet head (RJH). It is suggested that this situation is another where the myth of 'shear damage' has had a detrimental effect on the optimal operation of commercial bioprocessing.

See more of this Session: Mixing in the Biotechnology Industry
See more of this Group/Topical: North American Mixing Forum

Comparison of Traditional Rushton Disc Turbines with up-Pumping Hydrofoil B2 Impellers in 550 L Pilot Scale Aerobic Submerged Fermentations

Mads O. Albaek, Fermentation Pilot Plant, Novozymes A/S, 2880 Bagsværd, Denmark

Stuart Stocks, Novozymes A/S, 2880 Bagsværd, Denmark

Alvin W. Nienow, Chemical Engineering, University of Birmingham, Birmingham, United Kingdom

There has long been a debate concerning the most suitable impeller for aerobic submerged fermentation (and indeed, for gas liquid or 3-phase reactors in general). Much academic work evaluating agitator design with respect to flow pattern, power consumption in gassed and ungassed conditions, kLa, and mixing time has been published. In every single study, the conclusion is that an alternative design to the traditional Rushton Disc Turbine (RDT), offers some kind of advantage; mixer manufacturers have been quick to add such impellers to their portfolio. One configuration currently favoured is the use of multiple up-pumping, large diameter, low power number, high solidity ratio hydrofoil impellers because of axial flow, stable gas dispersion and power characteristics. Interestingly, this impeller design is almost as far as conceptually possible from the RDT.

A program initiated by Novozymes to evaluate the performance of twin up-pumping, high solidity ratio Hayward Tyler (previously APV) B2 impellers included gassed and ungassed power characterization in viscous media and with very high power draw as well as comparison with twin RDTs in pilot scale fermentors of 550 L with commercially exploited fed-batch A. oryzae fermentation expressing recombinant proteins. This work extended previous work with B2s and RDTs to higher power draws, gas rates and velocities that are relevant for modern submerged commercial fermentations at relevant scale.

The experimental design of this investigation was aimed at assessing the ability of the new impellers to improve the yield of productivity at a realistic scale; a most important issue which no studies previously have covered. By variation of agitation and aeration of the process it was possible to determine the influence of these parameters as well as viscosity on the mass transfer capacity, biomass concentration and product concentration. Thus a kLa prediction that includes both impeller types was determined for the system.

A well known problem that often arises when working with filamentous fungi is the resulting viscous fermentation broths, which greatly lowers the maximum rate of oxygen transfer and makes the control of dO2 concentration at that desired difficult. This study resulted in a rheological prediction of the fermentation broth viscosity derived from the power law based on the biomass concentration and the vessel energy dissipation/circulation function.

The results of the comparison of the RDT and B2 impeller performance in pilot scale will be presented. Furthermore using the viscosity and mass transfer prediction it was possible to model the entire fed-batch fermentation process and predict the outcome of varied rates of agitation and aeration. Examples of such simulations will be shown.

See more of this Session: Mixing in the Biotechnology Industry
See more of this Group/Topical: North American Mixing Forum

Performance Characterization and Scalability of Disposable Bioreactors

Ashraf Amanullah, Oceanside Process Research and Development, Genentech Inc., Oceanside, CA

Terrence Allotta, Oceanside Process Research and Development, Genentech Inc., Oceanside, CA

Kapil Kothary, Oceanside Process Research and Development, Genentech Inc., Oceanside, CA

Gayle E. Derfus, Oceanside Process Research and Development, Genentech Inc., Oceanside, CA

Thomas Myint, Oceanside Process Research and Development, Genentech Inc., Oceanside, CA

Terry Hudson, Oceanside Process Research and Development, Genentech Inc., Oceanside, CA

The availability of compact, flexible and modular biomanufacturing technologies using disposable technologies has enabled their widespread use for toxicology and clinical manufacturing. Such facilities not only have the advantage of smaller footprint compared to traditional stainless steel plants (e.g. SIP and CIP piping) but also offer benefits of modularity and flexibility and reduced build-out time. Disposable bioreactor platforms, when combined with high efficiency processes and downstream disposable technologies can transform the manufacturing paradigm. This presentation will describe the use of pilot scale disposable bioreactors (rocking platform as well as stirred tank) in cell culture process development applications, including mixing and mass transfer characterization, comparability to stainless steel bioreactor and scalability performance.

See more of this Session: Mixing in the Biotechnology Industry
See more of this Group/Topical: North American Mixing Forum

The Simplest Stirred Tank Bioreactor

Mario M. Alvarez, David Bulnes-Abundis, Gayathree Murugappan, Leydi Carrillo-Cocoom, Ernesto Aguirre-Ezkauriatza, Marisa Granados-Pastor, and Josefina Castillo-Reyna

Centro de Biotecnología FEMSA. Tecnológico de Monterrey.
Ave. Eugenio Garza Sada 2501 sur. Monterrey, N.L. México. C.P. 64849.

Abstract. Stirred tanks are the most popular bioreactor configuration used in the Biotechnology Industry. However, screening protocols at the lab scale (i.e. to find optimum culture media or initial cell density) mainly rely on Petri dishes, culture bottles, spinner flasks or Erlenmeyer flasks. There is an evident need of fully scalable cell culture mini-bioreactors for screening applications. We studied the mixing performance, and culture adequacy of a 30 ml eccentric stirred tank bioreactor. A detailed mixing characterization of the proposed bioreactor is presented. LIF experiments and CFD computations are used to identify operational conditions required for adequate mixing. Cell culture experiments are conducted to validate this novel bioreactor system for: (a) recombinant bacteria culture, (b) CHO cell culture, and (c) expansion of human hematopoietic progenitor cells. Our results suggest that the proposed bioreactor configuration could be particularly useful several biotech/biomedical applications ranging from screening of conditions for production of biopharmaceuticals, stem cells expansion and large scale culture of plant cells.

AIChE extended abstract

Key words: Cell culture, mixing, screening, biopharmaceutical, microbioreactor

INTRODUCTION

Stirred tanks are the most ubiquitous type of bioreactor geometry in industrial Biotechnology. Either operated in batch, fed-batch or continuous mode, stirred tanks have proved to be flexible and reliable to accommodate bacterial, yeast, plant cell and animal cell culture applications. For example, most biopharmaceuticals are recombinant glycoproteins secreted by mammalian cells cultured in large scale stirred tank bioreactors (Nienow, 2006), typically in the range from 50 to 20,000 L. On the other hand, Biotechnology industry relies intensively in screening applications. In order to define the best formulation for a culture media, the optimum temperature at which to run a culture process, or to identify the most productive strain from a pool of recombinant clones, a series of screening experiments is required. Commonly, the first stages of that screening process are conducted in very small volume systems, in vessel geometries that share nothing in common with industrial stirred tanks. Petri dishes, culture bottles of 20 to 50 ml and Erlenmeyer flasks continue to be the preferred containers to grow and select high producing cells at the lab scale. However, the hydrodynamic conditions in a Petri dish or an Erlenmeyer flask do not have much in common with the next scaling stage, typically a stirred tank bioreactor. There is an uncovered niche for screening systems, stirred

tank-like, that can partially substitute the activities now done in small scale lab culture devices and that can assure scalability in terms of flow field conditions.
In a bioreactor, cell growth is a complex process where biological performance depends on multiple factors such us oxygen mass transfer, temperature and pH control, nutrient availability, among others. In turn, most of these factors are highly influenced by mixing performance. To achieve good mixing conditions in small volumes is not a trivial task. Typical stirred tank geometries used in industrial or lab scale biotechnology normally rely on turbulence as their mixing mechanism. But at smaller scales, i.e. example screening bioreactor scales, creation of turbulence might pose practical problems. In a 3 cm diameter reactor, such as the one studied here, to achieve the same degree of turbulence than in a stirred tank of 10 cm in diameter operated at 600 RPM, we would need to agitate at 6667 RPM. This is not feasible or desirable. At this speed, the mechanical stability of the system would be questionable, and cell shear damage, particularly to shear sensitive cells such as mammalian or plant cells, would occur. Here we investigate what would be the simplest stirred tank bioreactor design to achieve good mixing (as required for biotechnology screening applications) at the lowest possible agitation speed. We propose a simple stirred tank system to be used as a bioreactor for screening cell culture applications. Its mixing performance and adequacy as a cell culture vessel are validated through experimental flow pattern visualization experiments, computational fluid dynamics simulations (CFD) and culture of different types of cells, namely aerobic bacteria and mammalian cells.

MATERIALS AND METHODS

Stirred tank system. The stirred tank geometry analyzed here is depicted in figure 1. A 30 ml glass round-bottom vessel was agitated using a disc impeller. The two main characteristic features of this system are: (a) eccentric location of the agitation axis, as previously suggested by Alvarez et al. (2002a) and Cervantes et al. (2006), and (b) a simple impeller disc (with no blades) inclined 45° with respect to the horizontal plane. The impeller (0.7 cm radius) was positioned at an eccentricity of E=0.33, where eccentricity (E) is defined as the distance between the shaft and the vertical centerline of the tank/tank radius.
Flow visualization experiments. All visualization experiments reported here were done using glycerin (viscosity equal to 1200 cP @ 22°C) as fluid. Mixing 3D patterns within the bioreactor were observed through fluorescent tracer injections in a dark room where UV light was the only illumination source. To better resolve the evolution of mixing structure in time, laser induced fluorescence experiments (LIF) were conducted. In an otherwise dark room, a plane of laser light was projected towards the experimental system, to visually expose the mixing structure at a vertical 2D plane (slightly behind the rotation axis) of the bioreactor originated by the injection of a pulse of rhodamine B dye (Sigma, USA). Photographic documentation of the visualization experiments was done using a professional Canon® Rebel XSi camera.
CFD simulations. Computational Fluid Dynamics (CFD) experiments were done using an HP xw8600 Workstation with 2 Xeon Quadcore processors. Geometries were built using SolidEdge® and meshes were built using software from ACUSIM®. Velocity fields were calculated by solution of the Navier Stokes Equations at 2,937,070 volume elements using ACUSOLVE from ACUSIM® Sofware Inc., Mountview Ca. The fluid was modeled primarily as a Newtonian fluid with the nominal properties of water. Visualization was by means of Kitware Paraview®

Cell culture experiments. Cell culture experiments in the proposed bioreactor design were performed with three different cell types: (a) a recombinant strain of *Escherichia coli* (Rosetta gami ™., Novagen®,USA); (b) a CHO cell strain adapted to suspended culture in chemically defined medium (CD OptiCHO™ Medium, GIBCO.USA) and; (c) human hematopoietic stem cells (CD34+) isolated from umbilical cord blood cultured in DMEM F-12 supplemented culture media [10% FBS, 1% penicillin-streptomycin solution, 10 mM HEPES]. *Escherichia coli* was cultured in LB media, according to standard protocols recommended elsewhere for the batch culture of recombinant Rosetta gami strains (Kane, 1995). Cells were incubated at 37 °C. No pH control was implemented. An air stream was injected over the reactor liquid surface. Samples were taken to monitor biomass concentration (by absorbance at nm) and pH (using an external pH electrode). CHO cells were cultured in chemically defined medium supplemented with glutamine at 37 °C, 200 RPM, with no air bubbling, no pH control and no culture medium addition after inoculation. In the case of the Human hematopoietic progenitor cells, 25 ml of cell suspension at a concentration of $1.75X10^4$ cells/ml was seeded in the tank, such that the ratio of vessel width to liquid height was 1:1. DMEM F-12 supplemented culture media [10% FBS , 1% penicillin-streptomycin solution, 10 mM HEPES] was changed every 24 hours according to a feed batch protocol, where agitation was suspended for 5 minutes to allow cells to settle in the bottom of the tank and 80% of the culture media was removed from the upper portion of the tank using a sterile syringe and replaced by fresh media warmed to 37°C.

RESULTS AND DISCUSSION

Ideally, a stirred tank for biotechnology screening applications should posses a very simple geometry. Yet, it should mix efficiently, preferably at low speeds (to make possible the culture of shear sensitive cells). While at high Reynolds flows turbulence is the mechanism to achieve mixing, in laminar conditions (low speed), chaos must be triggered to achieve some degree of mixing. However, to mix effectively in a laminar system, we must achieve globally chaotic conditions (Lamberto et al., 2000). The proposed stirred tank geometry relies on geometrical asymmetry to achieve globally chaotic conditions (and therefore good mixing) in a very simple geometry: an unbaffled round bottom tank with no air injection stirred with an inclined disc impeller eccentrically located. Alvarez et al (2002b) demonstrated that a disc impeller located eccentrically can generate chaotic behavior in laminar stirred tanks, but not necessarily mix efficiently. Indeed, axial impellers eccentrically located performed better, in terms of mixing efficiency, than disc impellers. Here we demonstrate that a 45° inclined impeller provides efficient, practically globally chaotic conditions, at low Reynolds numbers. We conducted a series of tracer mixing experiments to study mixing performance at different eccentricity values. Figure 1(a-d) shows 2D planes of the mixing structure observed at low Re values (Re~30, agitation=300 RPM in glycerin) at different stirred tank configurations. Glycerin is used here to retard the effect of diffusion, allowing for a cleaner observation of the structural effect of convective mixing at very low speed. In terms of Reynolds number values, and consequently mixing structure, these conditions will be approximately equivalent to those observed in water at 0.5 RPM. Figure 1a presents the structural features of the flow induced by a concentrically located disc impeller inclined 45° with respect to its horizontal plane. In concentric stirred tank configurations operated in laminar regime, chaos and regularity coexist (Lamberto et al., 1996; Alvarez et al., 2002a). Some areas are rapidly invaded by the fluorescent dye (chaotic zones). Others remain segregated with respect to the chaotic region, as evidenced by the fact that the dye has not penetrated them even after several minutes of agitation (regular regions). Then, chaos and regularity do coexist, and are homogeneous in the concentric case (Figure 1a) and the eccentricity case E=0.21 (Figure 1b). In the concentric system, the flow is highly symmetric (Figure 1a). Chaotic regions of adequate mixing coexist with areas of low mixing intensity and regular behavior (zones were the fluorescent dye has not yet penetrated), as previously reported for stirred tanks operated in laminar regime (Lamberto *et al.*, 1999; Álvarez, 2002; Álvarez et al., 2002; Cervantes *et al.*, 2006). Specifically, segregated regions can be observed above and below the impeller and at the top of the stirred tanks. In the eccentric cases, the flow structure becomes progressively asymmetric. The segregated regions above and below the impeller get distorted at E=0.21, and practically are eliminated at E=0.33. At eccentricity values of E=0.33, the flow is practically globally chaotic, meaning that no regular regions of low mixing survived. In addition, axial circulation was greatly improved with respect to the concentric scenario.

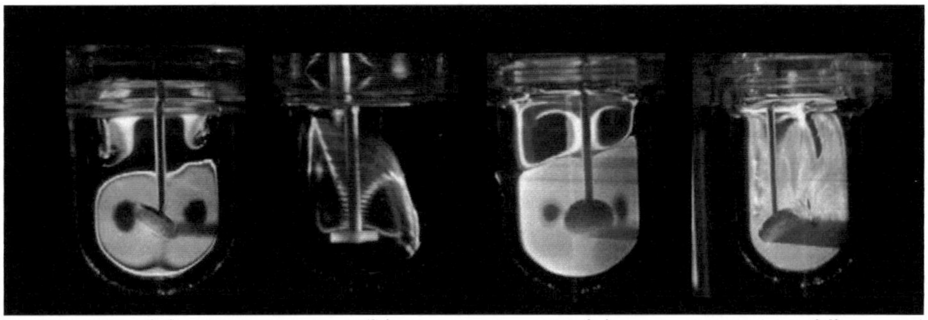

(a) (b) (c) (d)

Figure 1. Mixing visualization experiments: (a) laminar flow patterns are revealed through a fluorescent tracer injection under UV light in a concentric system (E=0) 10 minutes after injection, and (b) eccentric system (E=0.21) one minute after injection; (c) LIF experiments reveal the mixing structure of eccentric stirred tank systems agitated by horizontal flat disc impellers and (d) 45 angle disc impellers at low Re number.

As established previously (Alvarez et al., 2002b), in concentric conventional stirred tanks operated in the laminar regime, chaos is triggered by the periodic passing of the impeller blades, which perturbs the otherwise completely regular and symmetric manifold of the flow (Alvarez et al., 2002). Indeed, a concentric stirred tank with a disc impeller without blades will produce an entirely regular flow system, where convective mixing is not possible (Alvarez et al., 2003). In the absence of blades, a concentric system is completely axisymmetric, and therefore fully 2D. Chaos is not feasible in a 2D system. However, when the rotation axis of an impeller, even a simple disc impeller, is slightly shifted from the center line, the flow becomes a 3D system, and therefore susceptible to chaos. As previously shown by Alvarez et al. (2002), chaos is experimentally observable in flat bottom eccentric stirred tanks agitated by axial impellers, and even horizontal disc impellers, in the laminar regime. However, although capable of generating chaos, the mixing performance of eccentric horizontal disc impeller systems is perfectible. Cervantes *et al.* (2006) studied the mixing performance of a flat bottom eccentrically agitated laminar stirred tank. Using two different disc impellers with no blades differing in diameter and different rotational speeds, the authors estimated minimum mixing times in the order of ten minutes for a 1000 ml volume stirred tank. Here we demonstrate that inclined eccentric impellers can further improve mixing performance in eccentrically agitated tanks. Figure 2 shows CFD results from simulations done with two different disc impeller configurations (varying in angle) rotating at 150 RPM in a water-like liquid (matching average culture media properties) at an eccentric position (E=0.25) in a round bottom flask. A flat, horizontal disc, similar to the impeller geometry previously analyzed by Alvarez *et al.* (2002) and Cervantes *et al.* (2006), is compared versus a disc impeller, where the angle of the impeller with respect to its axis is different to zero (for the case reported here, 45 degrees). The effect of this small geometrical modification in the impeller angle is remarkable. In general, circulation patterns were dramatically modified, as shown in the velocity fields presented in Figure 2.

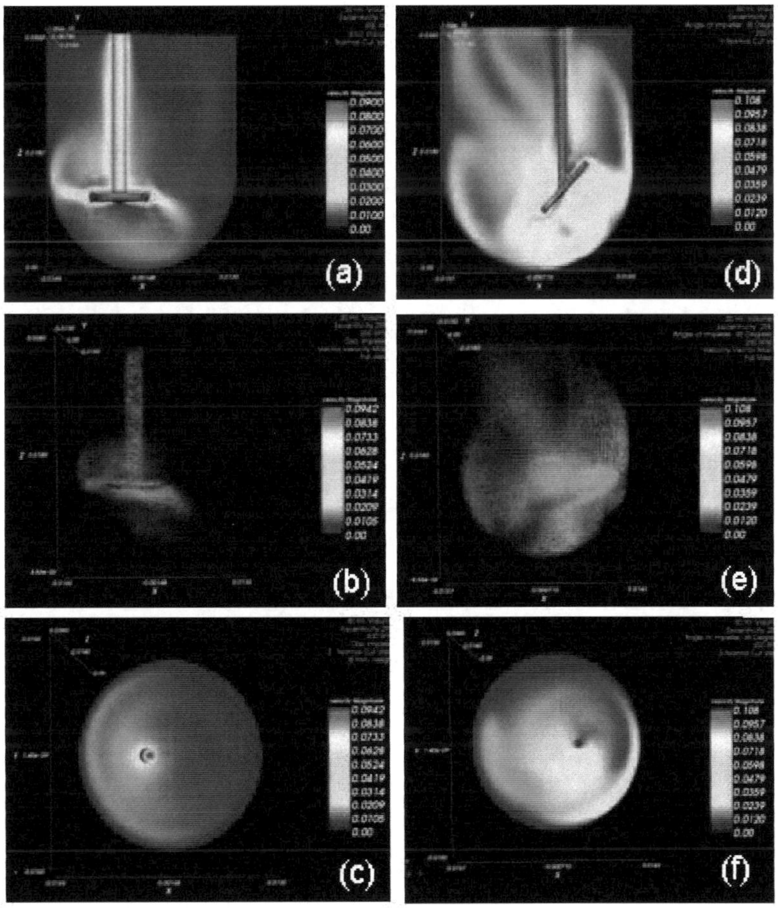

Figure 2. Velocity fields (obtained by CFD calculations) within eccentric stirred tanks (a) agitated by a horizontal disc impeller (vertical velocity contour at the axes level); (b) (vertical velocity field), and (c) (horizontal velocity contour at the impeller level); and (d) agitated by a 45° disc impeller (vertical velocity contour at the axis level); (e) (vertical velocity field), and (f) (horizontal velocity contour at the impeller level).

Figure 2(a-b) presents a comparison of the velocity fields corresponding to a vertical plane of eccentric stirred tanks only differing on the impeller inclination angle. In figure 2(e-f), a contrast of 3D velocity vectors is shown. At the same rotational speed, the impeller effect is noticeably wider in the case of the inclined impeller, as velocity fields corresponding to vertical (Figure 2(a-b)) and horizontal (Figure 2(c-d)) planes within the tank. Particularly, axial circulation was greatly improved by the use of an inclined disc impeller as illustrated by the 3D vector maps shown in Figure 2(e-f). Note that comparatively, higher magnitude vectors visit the bottom and the top of the tank agitated by the inclined disc. This is particularly important in the context of bioreactor engineering. Typically multiple phase systems, in stirred tank bioreactors the adequate suspension of cells depends on the bottom circulation patterns. In addition, mass transfer through the liquid surface is governed by circulation patterns at the top section of the tank. As we explain next, as the scale is reduced, oxygen transfer through the liquid surface becomes a preferred mechanism. To propitiate mixing and adequate mass transfer in a small scale stirred tank is not trivial, particularly for aerobic systems. Mixing and adequate oxygen mass transfer are not easily achievable in small scales by conventional bubble injection. In larger systems, agitation with marine turbines causes turbulence as the main mixing mechanism. Under turbulent conditions, oxygen dosed through fine spargers is correctly distributed, and typical bubble residence times allow for adequate oxygen mass transfer. As scale reduces, shorter residence time for air bubbles is expected, and adequate gas dispersion will demand higher shear rates. In addition, more bubbles per time unit will get to the surface of the tank, trapping cells at their surface. Both, high shear rates and bubble damage have been studied as possible factors of lethal and sub-lethal damage to mammalian cells, recognized as more fragile to forces than bacteria or yeast (Keane *et al.*, 2003; Amanullah *et al.*, 2004). This framework almost precludes the use of typical impeller designs and gas dispersion strategies for aerobic processes at very small scales. However, in typical stirred tank geometries, the liquid surface/volume ratio is relatively low compared to that of Petri dishes, culture bottles or stirred tanks. Therefore, axial circulation and liquid surface renewal should be improved to make aerobic cell culture feasible in small scale stirred tanks.

Figure 3 presents CFD velocity field calculations for 45 degree angle configurations at eccentric values E=0.33. Different velocity field contours (from 75 to 200 RPM) are compared. Although the underlying flow structure is similar among them, there are significant differences in the axial outreach of the flow field. As the rotation speed increases, the circulation at the upper section of the tank is improved. At 75 and 100 RPM speed values, the upper section of the tank exhibits low local speed values. Progressively, axial circulation is improved as speed increases. At 200 RPM, more active flow patterns are evident near the liquid surface, which assures higher mass transfer rates from the liquid surface.

To investigate if oxygen mass transfer rates form the liquid surfaces are enough to support aerobic cell growth, a recombinant *Escherichia coli* strain (Rossetta gami™ from Inovagen, Inc.USA), CHO cells adapted to suspension culture, and CD34+ human hematopoietic stem cells (from cord blood) were cultured at 200 RPM in a 30 ml culture media volume. As a sole form of oxygen supply, an air stream was continuously fed at a constant rate at above the liquid surface. The observed growth evolution of the *Escherichia coli* culture (see Figure 4a) exhibits a normal behavior. If a first order kinetic growth model is assumed ($dX/dt=\mu X$; where X is the biomass concentration at any given time, Xo is the initial concentration value, and t is time), then the specific growth rate (μ) during exponential phase can be calculated from the slope of the linearized and integrated biomass material balance: $\ln X= \mu t + \ln Xo$. The observed μ value for the experiments using our 30 ml screening bioreactor was of 0.66 h^{-1}, consistent with a value of 0.60 h^{-1} and 0.63 h^{-1} observed for the same strain and culture media in 150 ml Erlenmeyer flask experiments and in a stirred tank bioreactor.

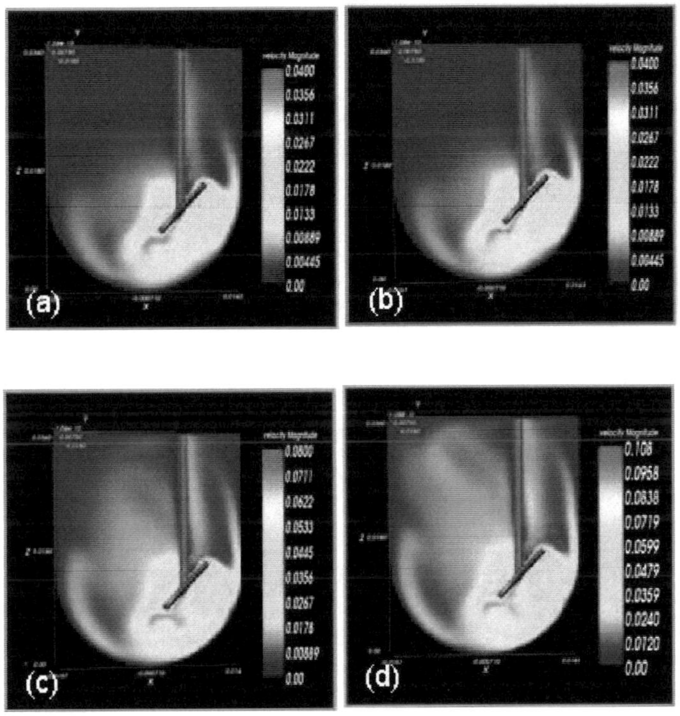

Figure 3. Velocity fields (obtained by CFD calculations) within eccentric stirred tanks agitated by a 45° disc impeller agitated at (a) 75 RPM; (b) 100 RPM; (c) 150 RPM; and (d) 200 RPM.

Similarly, in culture experiments with CHO cells, the specific growth rates observed during exponential phase are similar to those observed in conventional fully instrumented stirred tanks using the same cells and culture medium (μ=0.020 to 0.023 h^{-1}). Figure 4b shows the cell count and glucose time profiles for an experiment were $2.5X10^5$ cells were seeded in the 30 ml eccentric bioreactor operated at 200 RPM with no pH control. No air was bubbled to the tank, allowing oxygen mass transfer to occur exclusively through the gas/liquid interface. Cell densities higher than $1X10^6$ were achieved after 72 days of cultivation. This evolution and cell density were also similar to those observed during the same time period in fully instrumented bioreactors (data not shown).

In our experiments with hematopoietic stem cells, agitation was set at 200 RPM (in water-like media) to allow for adequate mixing while minimizing mechanical stress. Although adequate liquid-liquid mixing is achievable at much lower speeds, visualization experiments with neutrally buoyant 20 microns particles (results not presented here) demonstrate that higher speeds are needed to maintain cells in suspension. The "just suspension velocity" for polystyrene particles in water was estimated in 200 RPM.

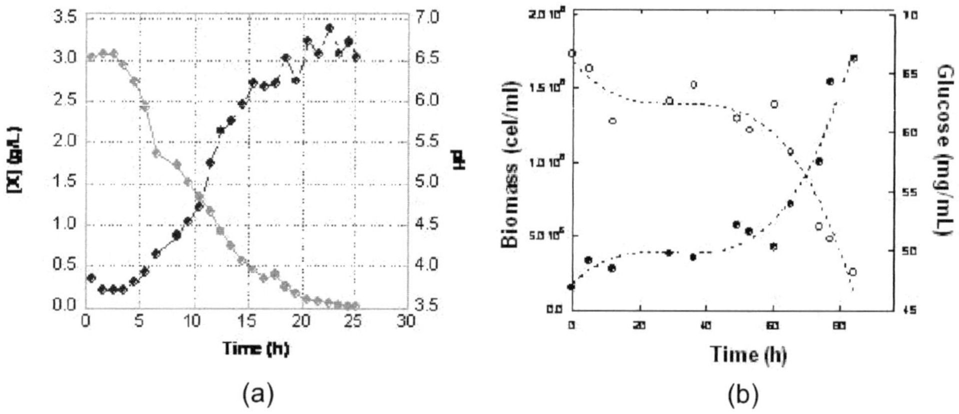

Figure 4. Cell culture experiments in a 30 ml eccentric stirred tank bioreactor with no air bubbling: (a) cell growth and pH profiles in time for the culture of a recombinant *E. coli* strain; (b) cell growth and glucose consumption profiles of CHO cells adapted to suspension conditions.

Cells cultured in the eccentric stirred tank bioreactor at an initial concentration of $1.75X10^4$ cells/ml exhibited growth curves (not shown), with characteristic latency, exponential and stationary phases. Cell doubling time, calculated using a linear regression analysis on the exponential growth phase, was 31.46 hours, 19.18% increase from doubling time in stationary culture in Petri dishes (standard culture protocol for this type of cells). Cell viability was maintained high (in the range of 80% to 100%). In exponential phase, cell viability values were sustained in the range of 95 to 100%, direct evidence that the flow field conditions are acceptable to favor survival and growth of extremely shear sensitive cells.

CONCLUSIONS

A 30 ml round bottom vessel stirred at relative low speed by an inclined disc impeller eccentrically located has been proposed for biopharmaceutical screening applications. A mixing analysis conducted using fluorescent tracer experiments under UV light, LIF visualizations and CFD simulations shows adequate mixing performance even at low laminar Reynolds number values (Re<100). The asymmetry of the geometry translates in an asymmetric flow field where globally chaotic mixing is observed at eccentricity values of E=0.33. The just suspended velocity in water was calculated in 100 RPM. Based on this analysis, operational conditions were decided for the culture of aerobic cells relevant to the biopharmaceutical/biomedical context: (a) recombinant *E. coli* cells, (b) CHO cells adapted to grow in suspended conditions, and (b) human non-embryonic hematopoietic stem cells. Escherichia coli cultures exhibited a typical growth behavior, with specific growth rates and maximum cell density values similar to those observed in Erlenmeyer flasks. CHO cells observed adequate cell growth and glucose consumption rates during the first 72 hours of culture with similar specific growth rates to those observed in instrumented stirred tank bioreactors. The stem cell culture was monitored over a period of 10 days, showing adequate behavior in terms of growth rates, cell densities and viability. The bioreactor design proposed here might be an alternative for small scale relative high through-put screening of biopharmaceutical/biomedical applications in aerobic stirred tank geometries.

REFERENCES

1. Álvarez MM. 2000. Using spatiotemporal asymmetry to enhance mixing in chaotic flows: from maps to stirred tanks. Ph. D. Thesis, Chemical and Biochemical Engineering. Rutgers University, New Brunswick, N.J.

2. Álvarez MM., Arratia PE., Muzzio FJ. 2002. Laminar Mixing in eccentric stirred tank systems. *Canadian Journal of Chemical Engineering, 80, 546.*

3. Amanullah A, Buckland BC, and Nienow AW. Mixing in the Fermentation and Cell Culture Industries in Handbook of Industrial Mixing, Science and Practice, edited by Paul EL, Atiemo-Obeng, V and Kresta SM. Willey Interscience. United States of America, 2004, pag1071-1170.

4. Cervantes MIS, Lacombe J, Muzzio FJ, and Alvarez MM. 2006. Novel bioreactor design for the culture of suspended mammalian cells. Part 1: Mixing characterization. *Chemical Engineering Science*, 61(24): 8075-8084.

5. Keane JT, Ryan D, and Gray PP. 2003. Effect of shear stress on expression of a recombinant protein by Chinese hamster ovary cells. *Biotechnology And Bioengineering* , 81(2): 211-220.

6. Lamberto DJ, Álvarez, MM, Muzzio FJ. 1999. Experimental and computational investigation of the laminar flow structure in a stirred tank. *Chemical Engineering Science, 54, 919.*

7. Nienow AW. 2006. Reactor engineering in large scale animal cell culture. *Cytotechnology,* 50(1-3): 9-33.

8. Kane JF. 1995. Effects of rare codon clusters on high-level expression of heterologous proteins in *Escherichia coli. Current Opinion in Biotechnology*. 6(5): 494-500.

Agitation of Microbial Cultures Via Low-Frequency Acoustic Energy for Improved Oxygen Transfer

Todd A. McAdams, Resodyn Corporation, Butte, MT

Conventional biological orbital shakers have substantial drawbacks that include low volumetric capacities, limited oxygen transfer capability and difficulty in mixing viscous liquid cultures. A novel mixing technology has been developed to overcome these limitations by applying low-frequency, high intensity acoustic energy to agitate liquid cultures. This new technology has been incorporated into a new incubator system offered by Applikon Biotechnology. This advanced technology is a significant improvement over orbital shaker/incubator devices, as it has been shown to dramatically enhance oxygen transfer, as well as substantially improve bulk mixing for microbial cultures. Combined with the novel air-pumping (Oxy-PumpTM, Applikon Biotechnology, Inc.) stopper, oxygen transfer rates up to 6-fold higher than orbital-shaken cultures can be achieved using the technology. Studies comparing the oxygen mass transfer coefficients between this technology and conventional orbital shaker technology will be presented. Various factors that influence the degree of oxygen transfer, including stopper design, intensity of agitation, vessel geometry, volumetric loading, and antifoam inclusion will be discussed. Culture experiments utilizing an E. coli HB101 strain engineered to express Green Fluorescent Protein (GFP) were used to compare orbital shakers (400rpm) to the acoustic mixing technology and peak biomass levels and GFP expression levels were found to be 4-fold and 14-fold higher, respectively, using the acoustic mixing technology. Additional studies using Pseudomonas fluorescens, Bacillus subtilis, Aspergillus oryzae, and Penicillium raistrickii have all shown enhanced biomass levels using the acoustic mixing system. This new technology has substantial potential to enhance the productivity of a wide range of microbial culture systems.

See more of this Session: Mixing in the Biotechnology Industry
See more of this Group/Topical: North American Mixing Forum

Inline Mixing Devices for Medical Applications

Y.Delmotte, A.Blom & A.Douteur
Baxter R&D Europe
BioScience – BioSurgery

Contact : yves_delmotte@baxter.com

Agenda

- The challenge : mixing fibrinogen & thrombin
- Using sintered porous polymer as a mixer
- Biological testing to evaluate quality of fibrin
- Assessment of mixer performance with Laser technique
- Testing of the medical device efficiency
- Why does it work? Topology study of the mixer
- Medical & non medical applications
- Conclusion

Fibrin Formation

Tisseel fibrin sealant simulates the last stage of the clotting cascade

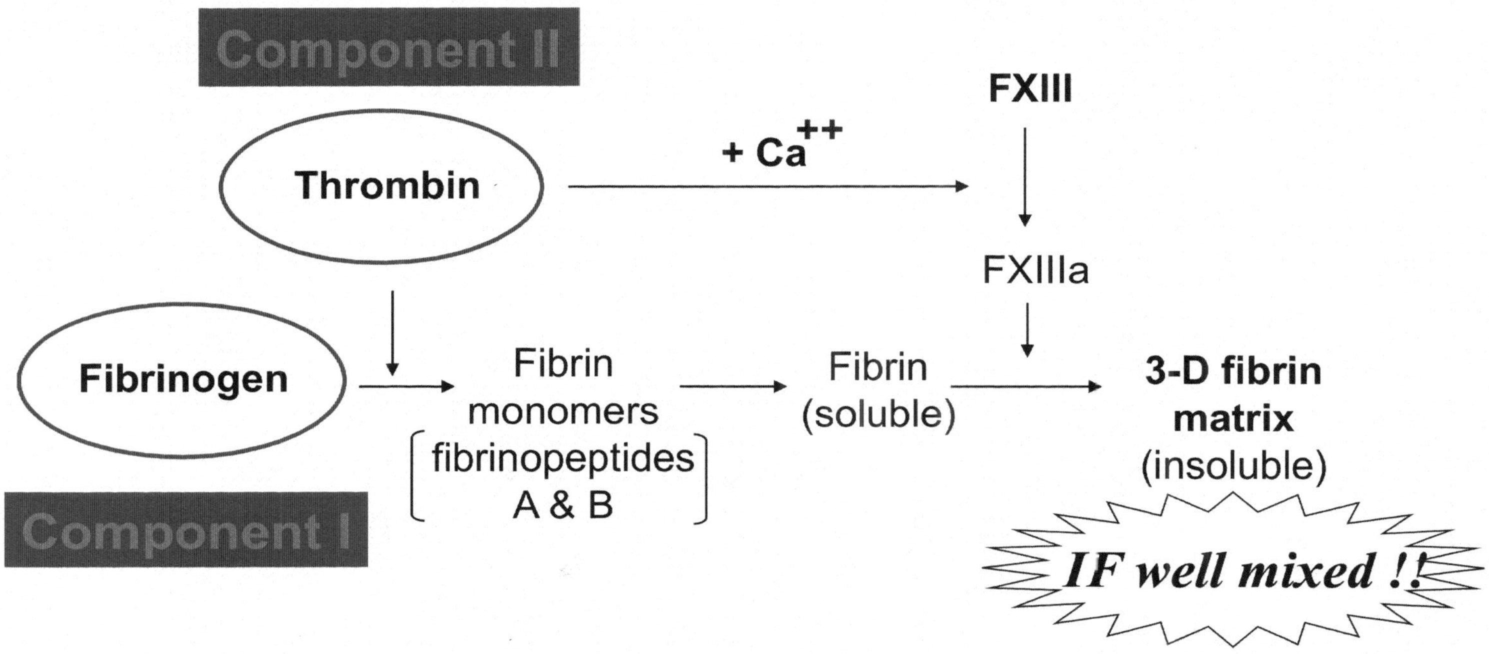

Fibrin : A versatile matrix

Structures and properties of the fibrin matrix can be controlled by modifying:

- Fibrinogen/thrombin concentration
- Ionic strength
- Calcium concentration
- pH
- Temperature
- Protein content
- Application methods: compression, lyophilization,...
- **Design of the mixing device**

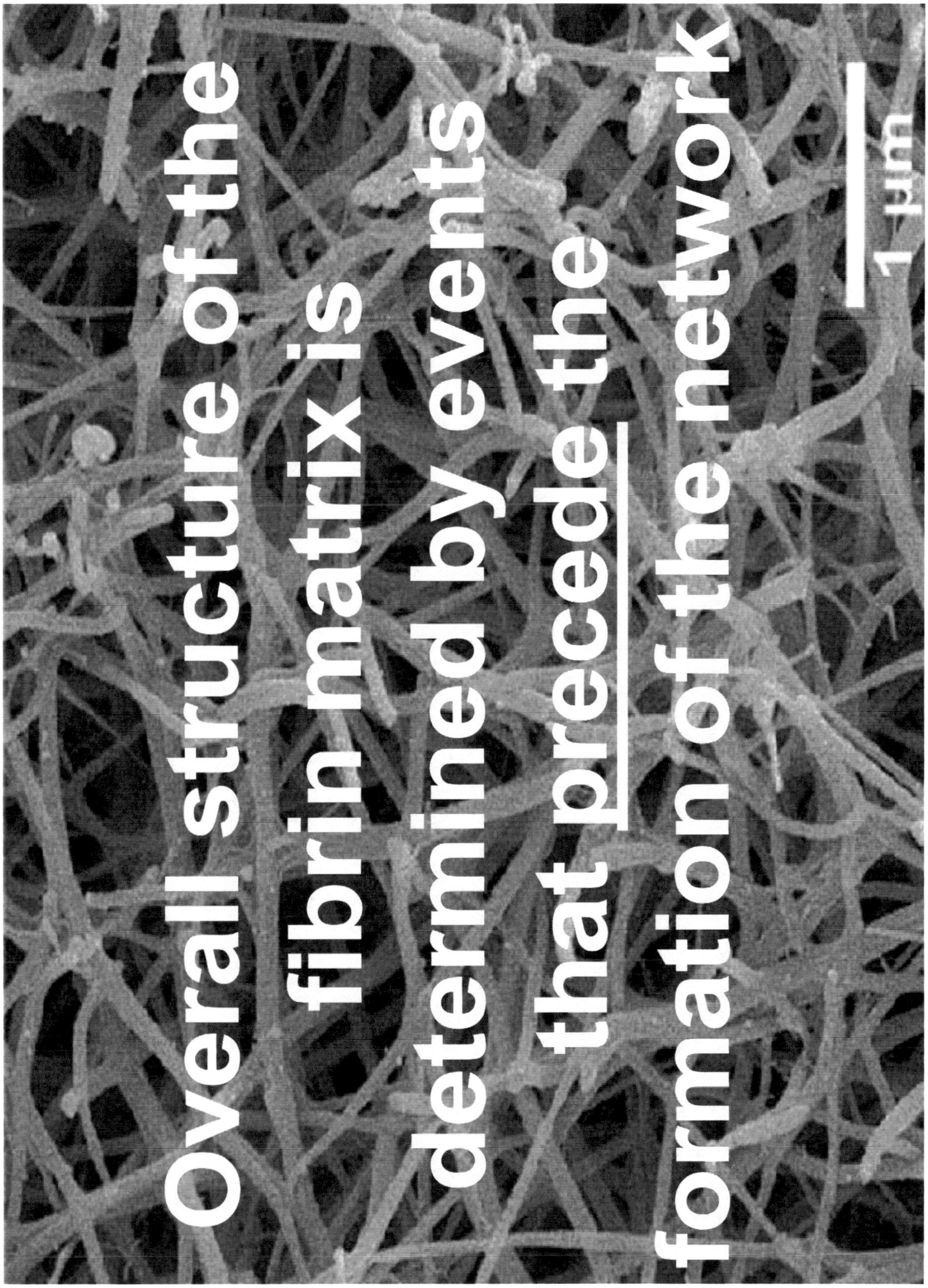
Overall structure of the
fibrin matrix is
determined by events
that precede the
formation of the network
1 µm

Technical constraints ....

<u>Mixing of Fibrinogen & Thrombin is difficult</u>:

- *Difference in viscosity*
- *Temperature dependent*
- *Flow rate dependent*

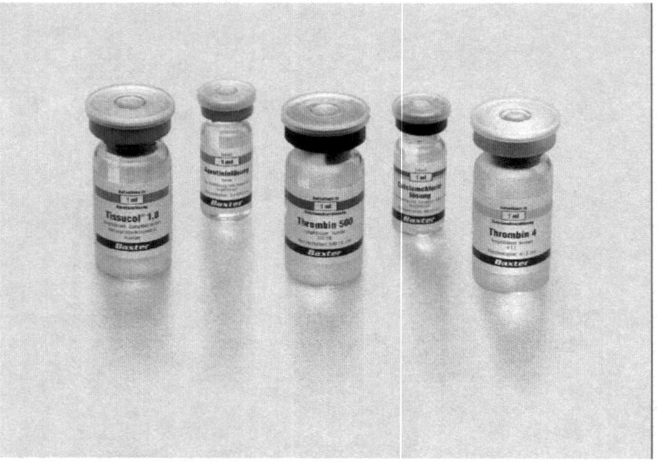

TISSEEL/TISSUCOL
Fibrin Sealant

Agenda

- The challenge : mixing fibrinogen & thrombin
- Using sintered porous polymer as a mixer
- Biological testing to evaluate quality of fibrin
- Assessment of mixer performance with Laser technique
- Testing of the medical device efficiency
- Why does it work? Topology study of the mixer
- Medical & non medical applications
- Conclusion

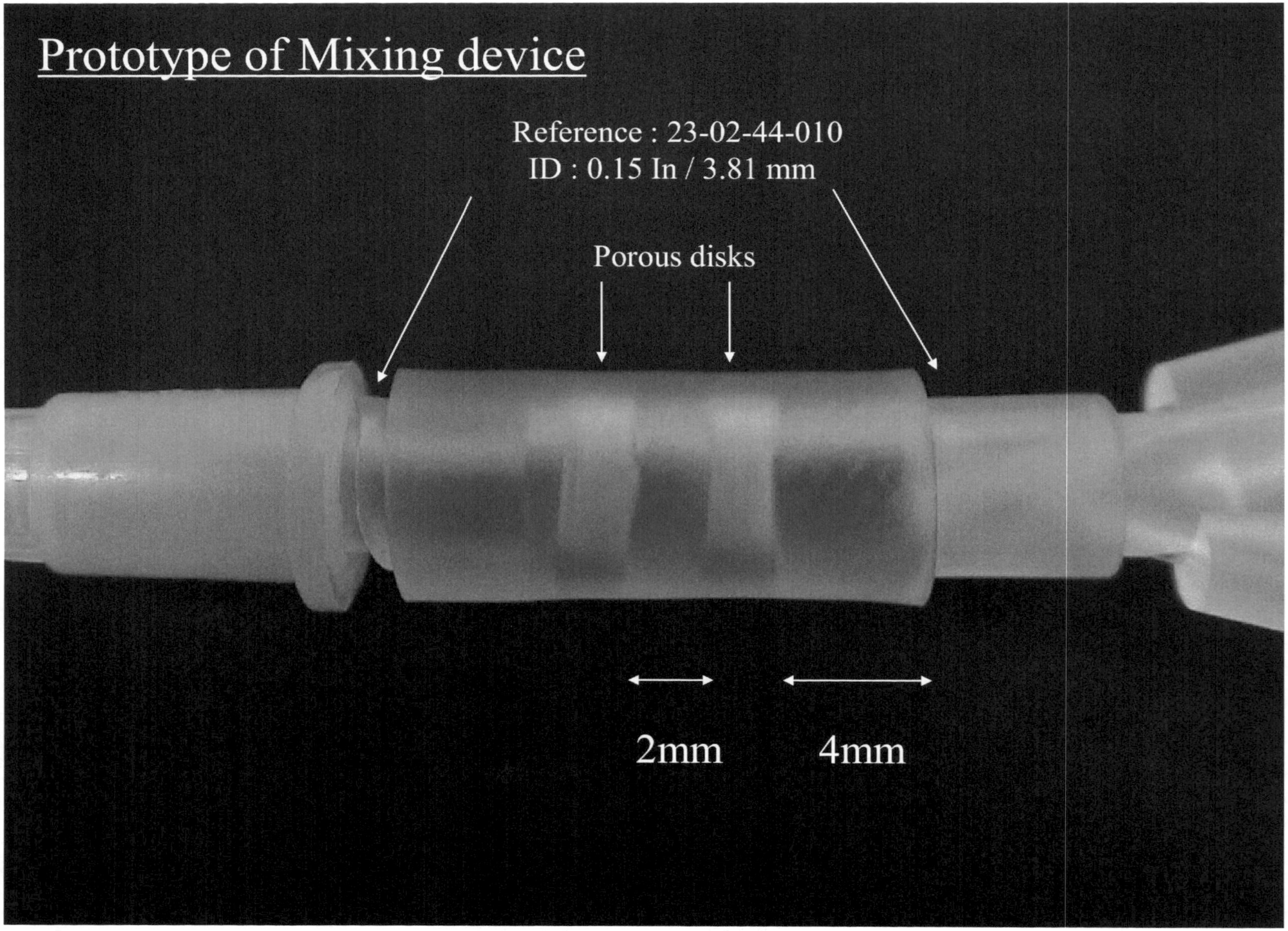
Prototype of Mixing device
Reference : 23-02-44-010
ID : 0.15 In / 3.81 mm
Porous disks
2mm
4mm

Agenda

- The challenge : mixing fibrinogen & thrombin
- Using sintered porous polymer as a mixer
- Biological testing to evaluate quality of fibrin
- Assessment of mixer performance with Laser technique
- Testing of the medical device efficiency
- Why does it work? Topology study of the mixer
- Medical & non medical applications
- Conclusion

Extensive testing campaign to evaluate the impact of variables on the *device design* and quality of *mixing*:

- Polymer (PE/PP, powder/grinded, suppliers…)
  - Porosity, thickness, PM (particulate matter)
  - Cutting process (Laser, rotary cut, molded)
- Number of discs in device : 1 vs 2 discs
- Distance between Y-piece and first disc
- Distance between discs
- Quality of mixing (Fibrin structure by electrophoresis, « Schnitzel test », turbidimetry,…)
- Effect of flow rate, T° on quality of fibrin formed
- Toxicology, Biocompatibility, USP & EP compliance…

Devices with 2 disks, VYON F, PE standard 1.5

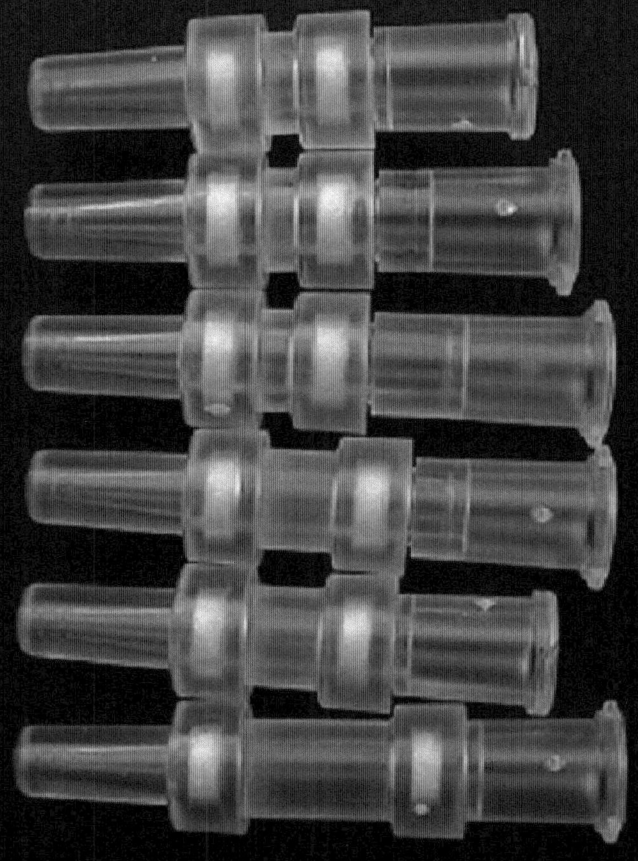

type : d-d y-d	Legend
7# : 4 - 2	PE std 1.5 2-2-4
2a : 4 - 4	PE std 1.5 2-4-4
1a : 4 - 6	PE std 1.5 2-6-4
3a : 6 - 4	PE std 1.5 2-4-6
4a : 6 - 2	PE std 1.5 2-2-6
5a : 10 - 2	PE std 1.5 2-2-10

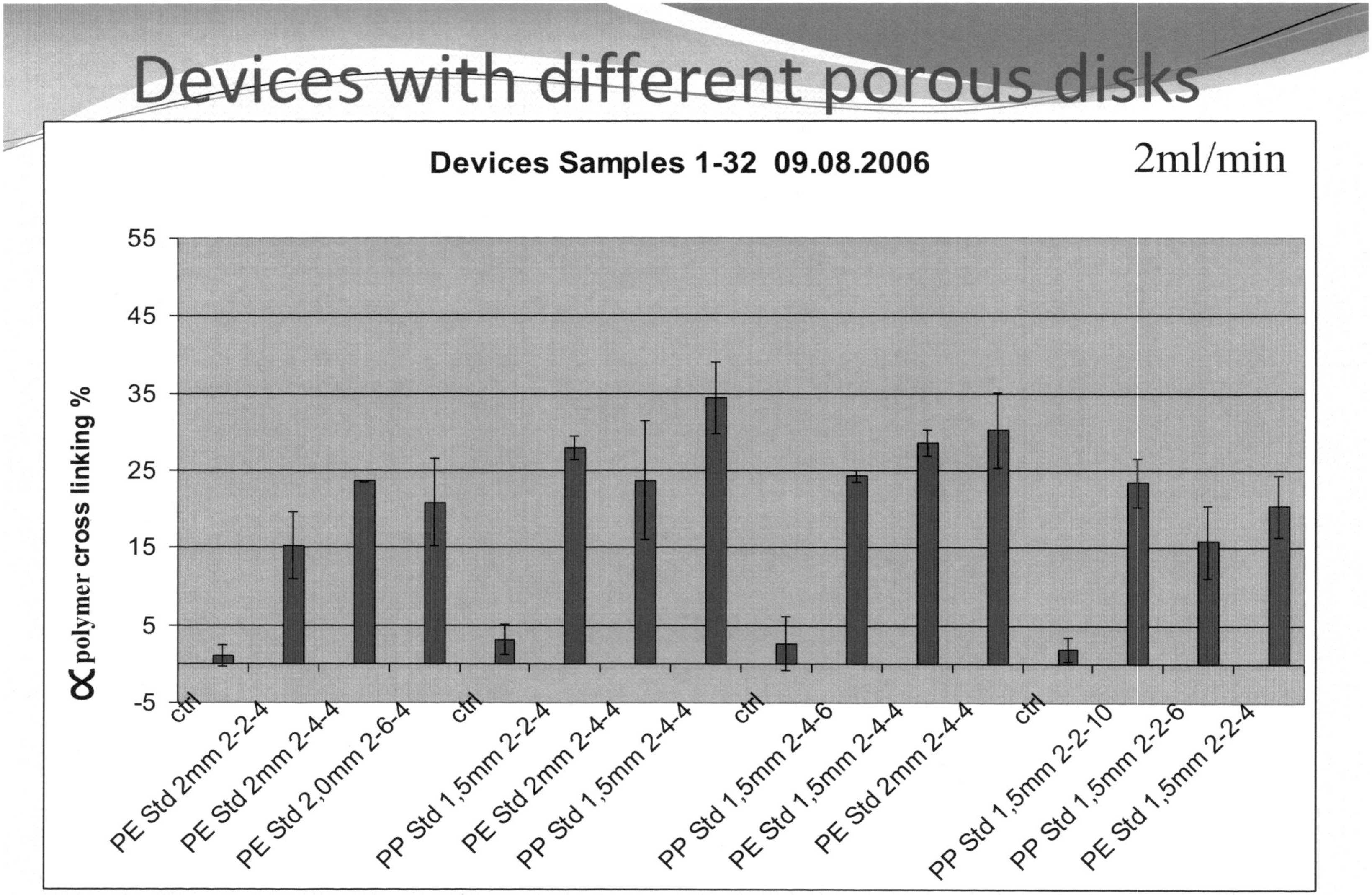

Devices with different porous disks
Devices Samples 1-32 09.08.2006
2ml/min
∝ polymer cross linking %
55
45
35
25
15
5
-5
ctrl
PE Std 2mm 2-2-4
PE Std 2mm 2-4-4
PE Std 2,0mm 2-6-4
ctrl
PP Std 1,5mm 2-2-4
PE Std 2mm 2-4-4
PP Std 1,5mm 2-4-4
ctrl
PP Std 1,5mm 2-4-6
PE Std 1,5mm 2-4-4
PE Std 2mm 2-4-4
ctrl
PP Std 1,5mm 2-2-10
PP Std 1,5mm 2-2-6
PE Std 1,5mm 2-2-4

Influence of temperature on mixing

2ml/min

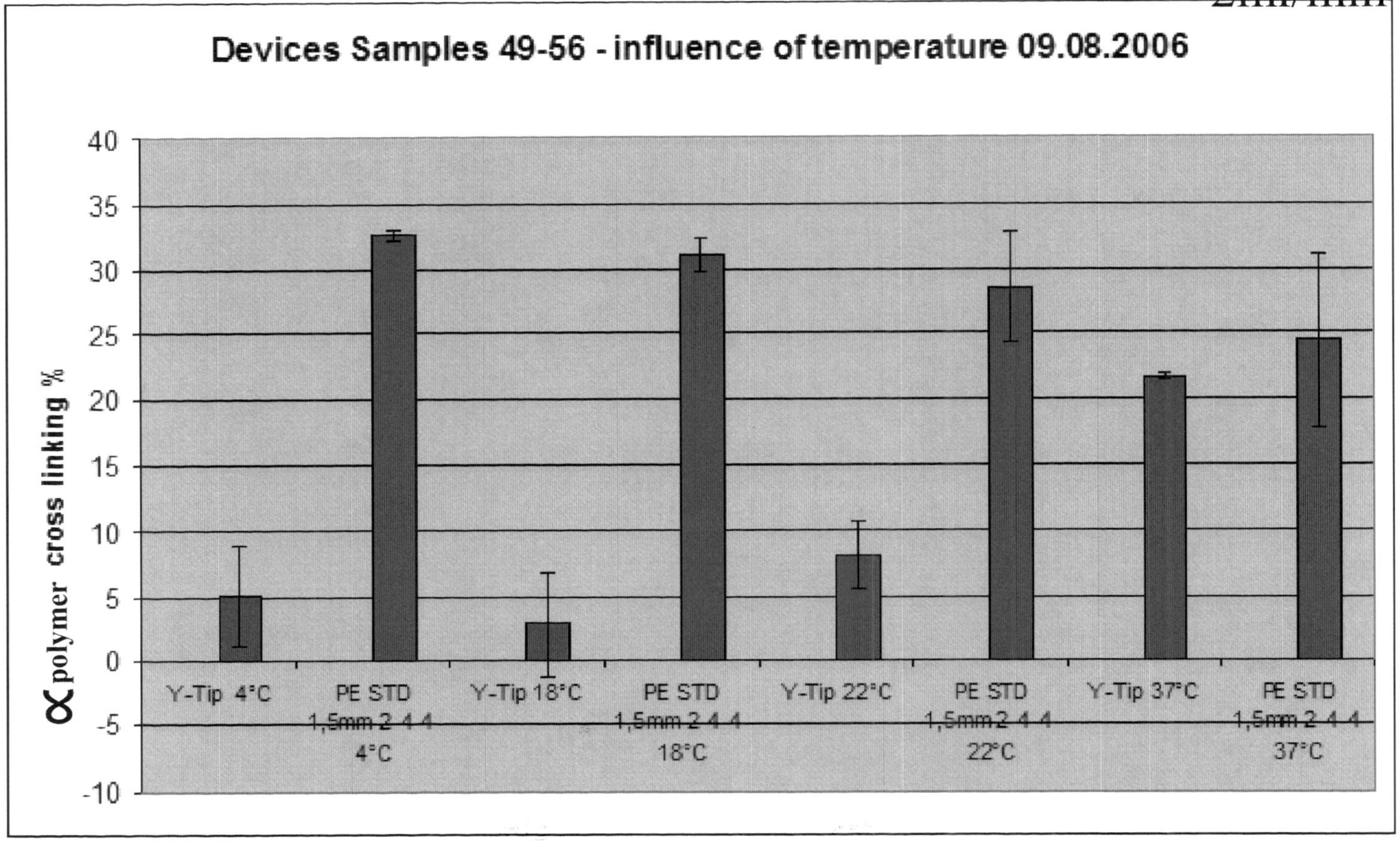

Influence of flow rate on mixing

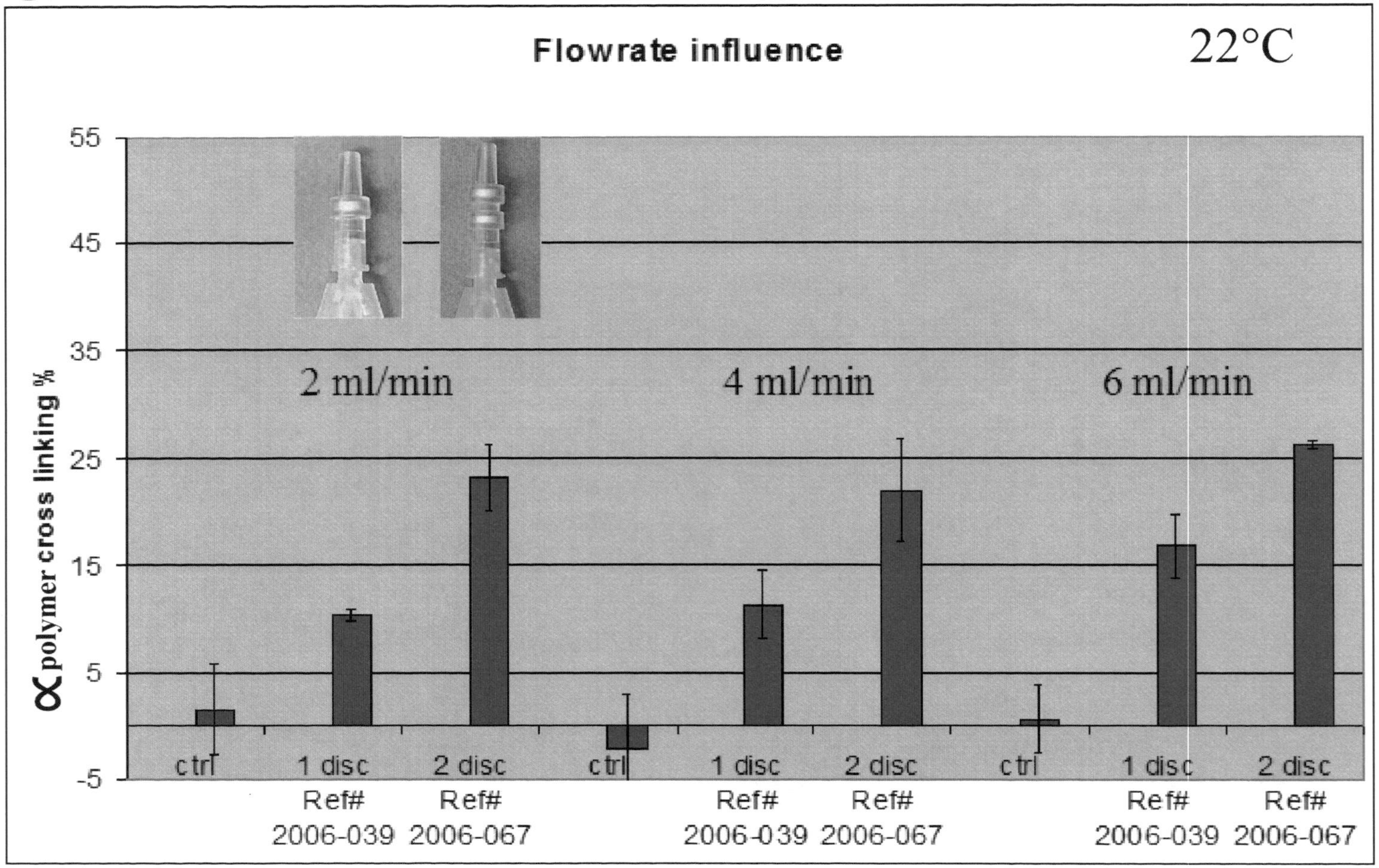

Agenda

- The challenge : mixing fibrinogen & thrombin
- Using sintered porous polymer as a mixer
- Biological testing to evaluate quality of fibrin
- Assessment of mixer performance with Laser technique
- Testing of the medical device efficiency
- Why does it work? Topology study of the mixer
- Medical & non medical applications
- Conclusion

Laser technology

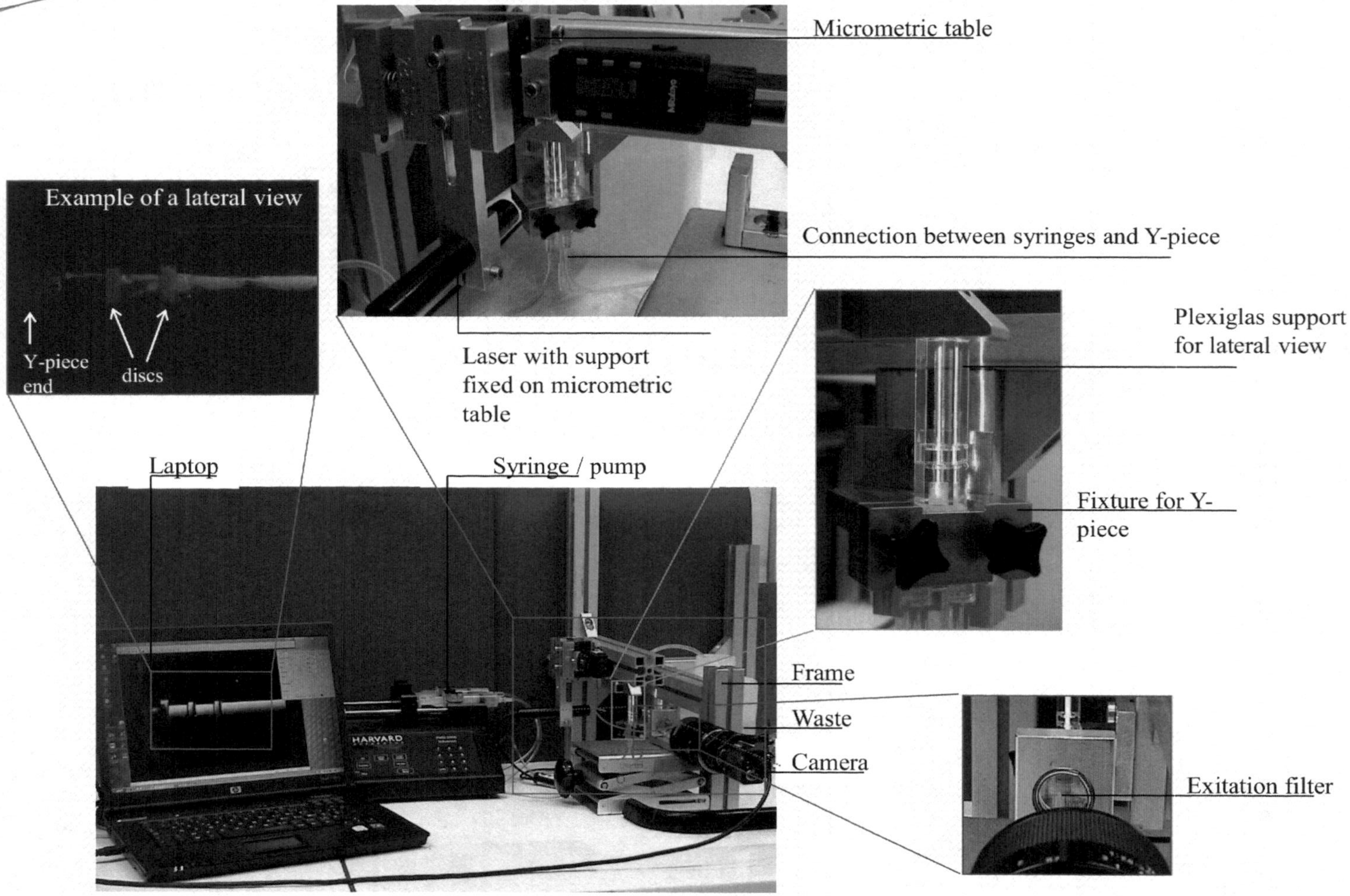

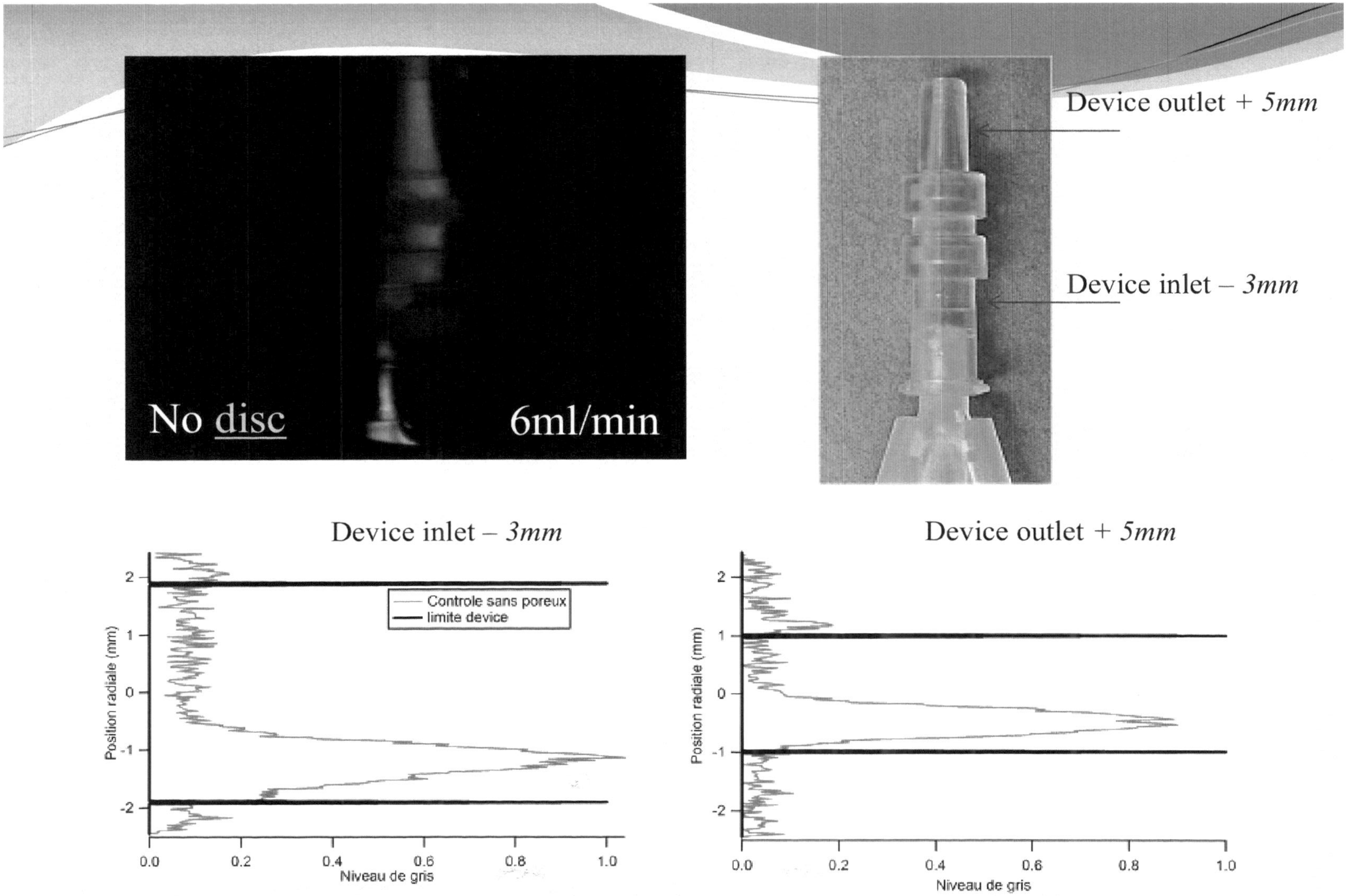
No disc
6ml/min
Device outlet + 5mm
Device inlet – 3mm
Device inlet – 3mm
Device outlet + 5mm
Controle sans poreux
limite device
Position radiale (mm)
Niveau de gris
Position radiale (mm)
Niveau de gris

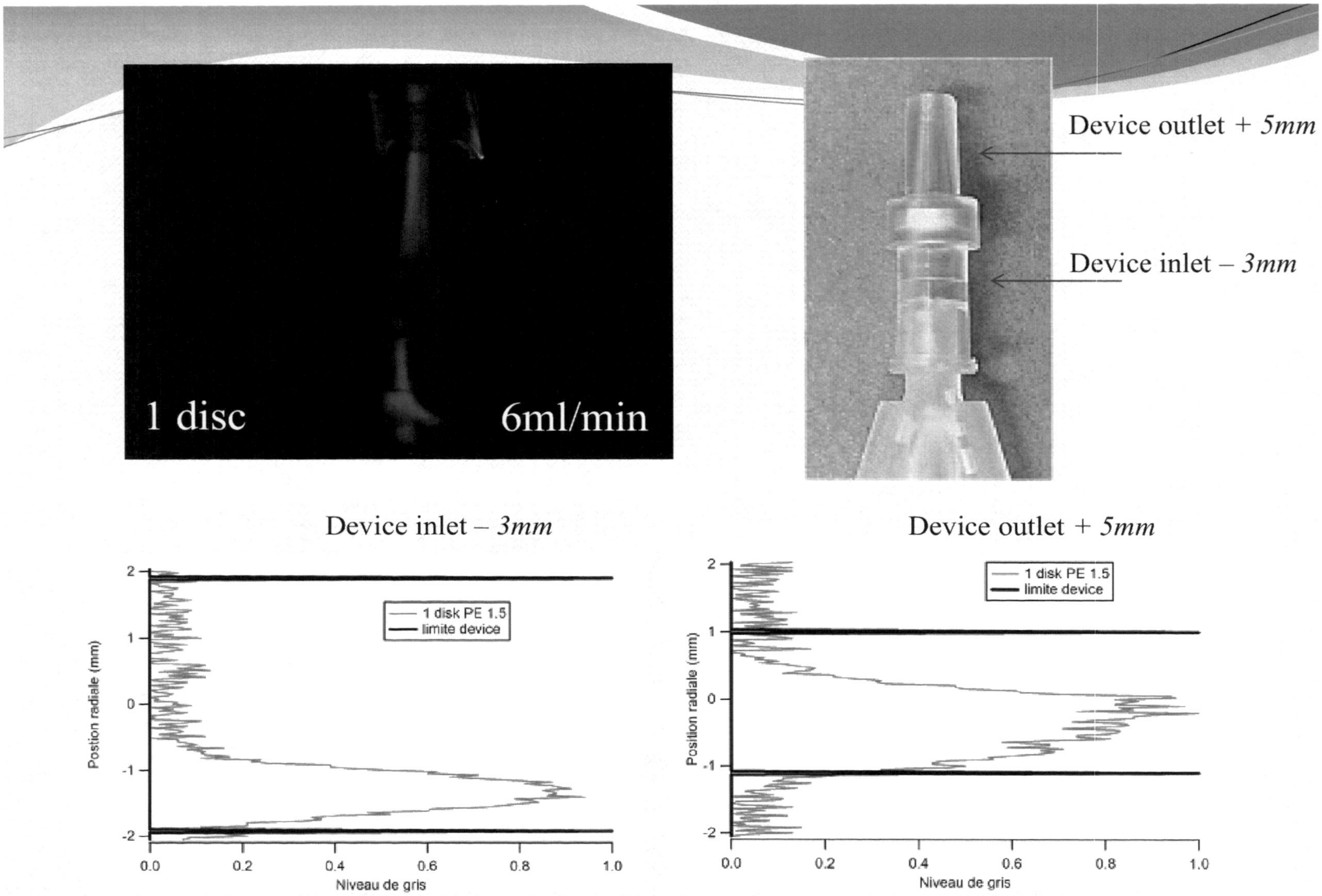
1 disc
6ml/min
Device outlet + 5mm
Device inlet – 3mm
Device inlet – 3mm
Device outlet + 5mm
1 disk PE 1.5
limite device
Position radiale (mm)
Niveau de gris

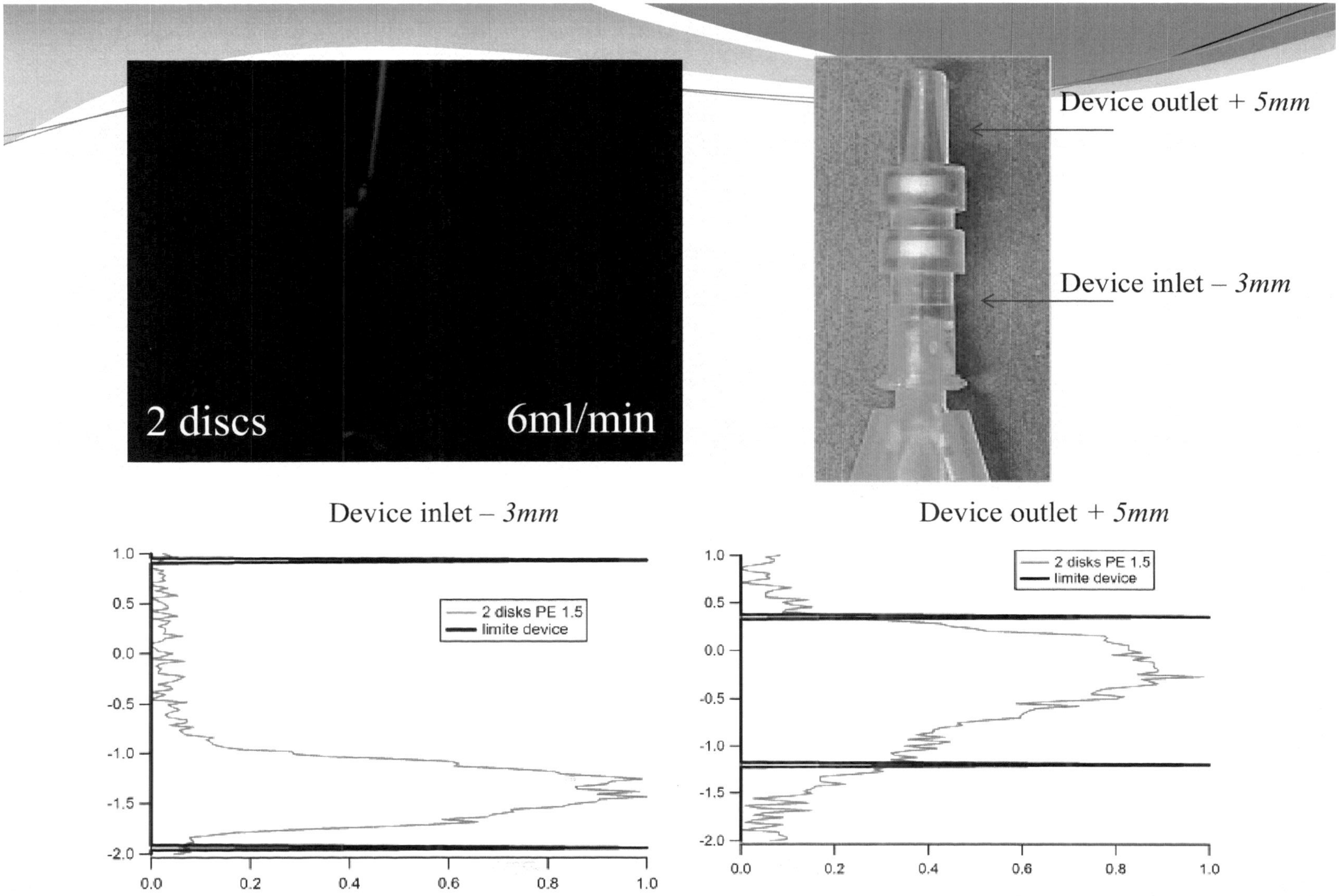
2 discs
6ml/min
Device outlet + *5mm*
Device inlet – *3mm*
Device inlet – *3mm*
Device outlet + *5mm*
2 disks PE 1.5
limite device
1.0
0.5
0.0
-0.5
-1.0
-1.5
-2.0
0.0
0.2
0.4
0.6
0.8
1.0

Agenda

- The challenge : mixing fibrinogen & thrombin
- Using sintered porous polymer as a mixer
- Biological testing to evaluate quality of fibrin
- Assessment of mixer performance with Laser technique
- Testing of the medical device efficiency
- Why does it work? Topology study of the mixer
- Medical & non medical applications
- Conclusion

MIXING OF ARTISS WITH VYON F 1.5MM MIXING DEVICE COMPARED TO CURRENT Y-TIP AND CANULA (SC0272008JG+SC0282008JG)

Absorbance Area Scan method:

•Application of the material into a 12-well plate
•Photometrically measurement with the Synergy2 MPR photometric system.

Example:

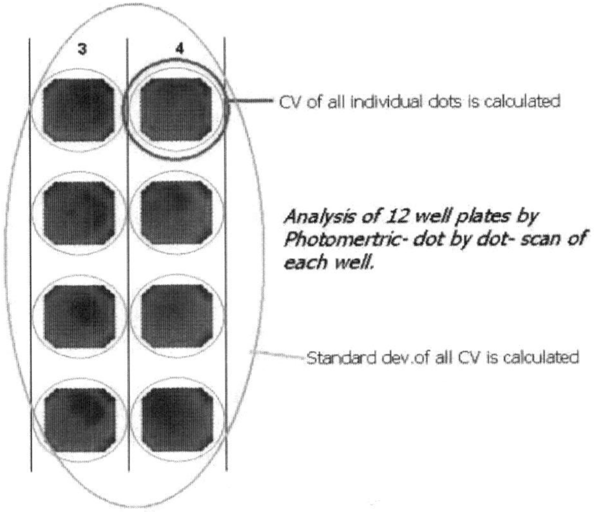

Coefficient of Variation (CV) is calculated as a measure of „degree of mixing":

Low CV corresponds to better mixing !

The Absorbance Area Scan methodology which can be applied using the Synergy 2 Microplate reader photometrically measures the absorbance of a defined number of sub-areas that together form the total area of one clot (= one well). For each individual sub area the turbidity is determined. As a measure of homogeneity of the whole clot the "coefficient of variance" (CV) is calculated for the sum of all data points covering one clot. Good mixing will results in low CV-values, sub-optimal mixing can be detected by an increase of the CV.

MIXING OF ARTISS WITH VYON F 1.5MM MIXING DEVICE COMPARED TO CURRENT Y-TIP AND CANULA *(SC0272008JG+SC0282008JG)*

Baxter

improved mixing tip (Vyon F) at (18°C) ⟷ comparison ⟷ **Current application Device at (33°C)**

<u>Comparison:</u> Y-piece / canula +disc (VyonF) (18°C) with Y-piece /current application needle (33°C) (three different Lots)

This figure shows the CV indicating degree of mixing – No significant difference (p= 0,173) between Y-piece / canula + disc (VyonF) (18°C) and the Y-piece /current application needle (33°C) could be detected. The Cyst Mix Device was used as a control. (note: each bar consists 18 data points, statistical analysis: One way ANOVA)

„Product Mixing at 18°C with new mixing Device is as least as good as current application at 33°C"

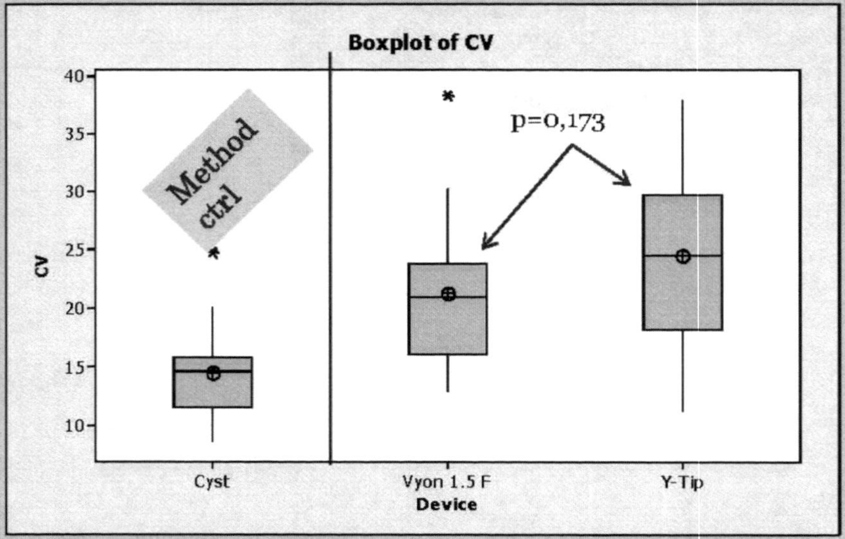

MIXING OF ARTISS WITH IMPROVED MIXING DEVICE COMPARED TO CURRENT Y-TIP AND CANULA (SC0272008JG+SC0282008JG)

Baxter

improved mixing tip (Vyon F) at (18°C) ⟷ comparison ⟷ **Current application Device at (18°C)**

Results:

Comparison: Y-piece / canula +disc (VyonF) (18°C) with Y-piece / current application needle (18°C) (three different Lots)

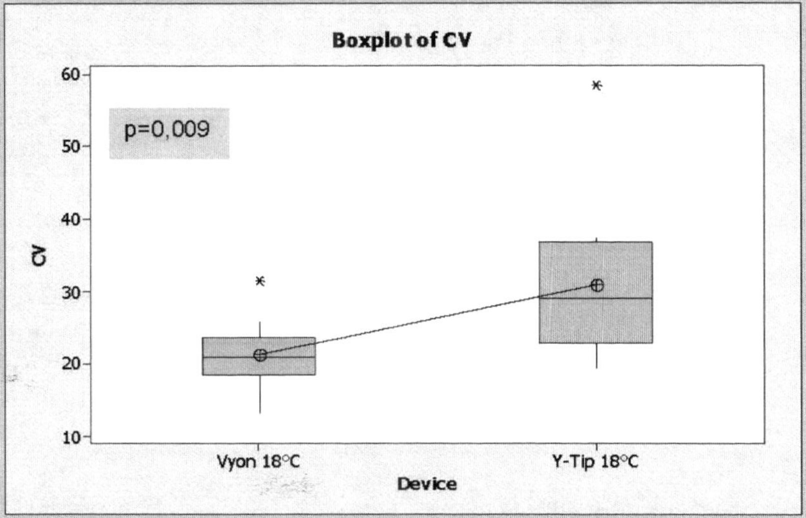

This figure shows the CV indicating degree of mixing – A highly significant influence of the device (p= 0,009) between Y-piece / canula +disc (VyonF) (18°C)and the Y-piece / current application needle (18°C) could be seen. (note: each bar consists 12 data points, statistical analysis: One way ANOVA).

"Improved mixing disc can better mix the product components (18°C)"

Agenda

- The challenge : mixing fibrinogen & thrombin
- Using sintered porous polymer as a mixer
- Biological testing to evaluate quality of fibrin
- Assessment of mixer performance with Laser technique
- Testing of the medical device efficiency
- Why does it work? Topology study of the mixer
- Medical & non medical applications
- Conclusion

Morphological study

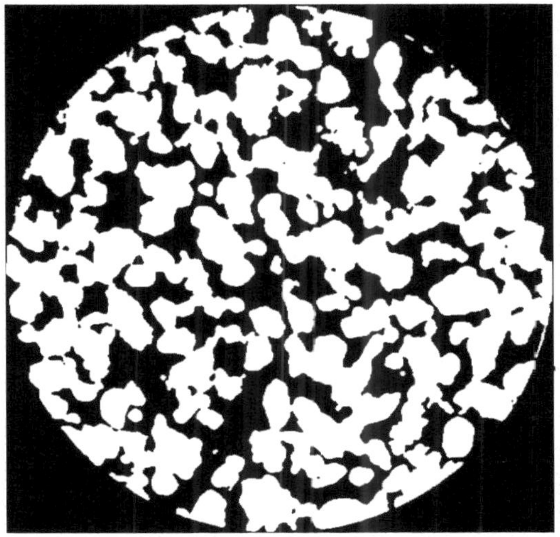

Slice of of VYON-F
after binarisation

- 3D High Resolution Tomography of VYON-F done at the Synchrotron, France
  → *Tomography (layer of 0.7 & 1.4 µm) of the disk used in our devices*
- File and data transfered to « École Polytechnique,Marseille, France »
- 3D Reconstruction of the data from tomography in order to study the morphology of the porous material to understand what are the parameters that are contributing to mix.

Here under is a picture of a slice of the porous material, by piling them up, they can reconstruct the whole 3D structure of the porous.

As 3D structure of an element of the porous Vyon-F disk, is reconstructed.
The following parameters have been analyzed

Volume of the cells

- liquid phase
- solid phase

Measure of the cols of the porous material

- number of pores surrounding each pore (connectivity)
- number of solid element surrounding a single solid element

Anisotropy

- number of intersections in axis X & Z

Volume of the cells

- liquid phase
- solid phase

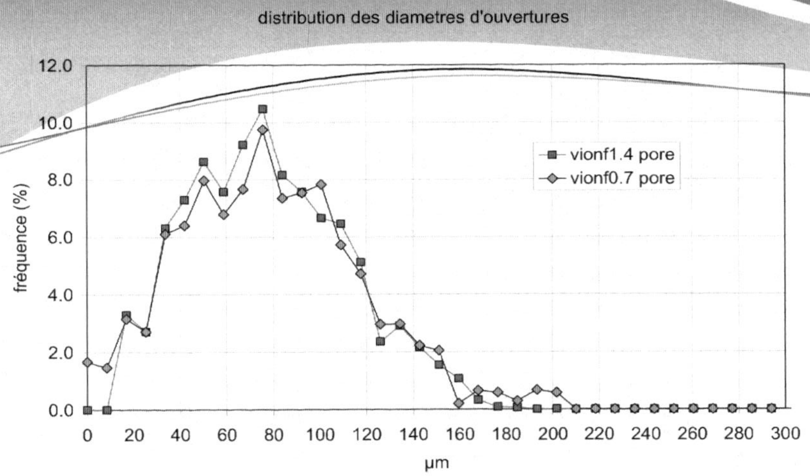

Distribution of the cell volume (pore size) of liquid phase

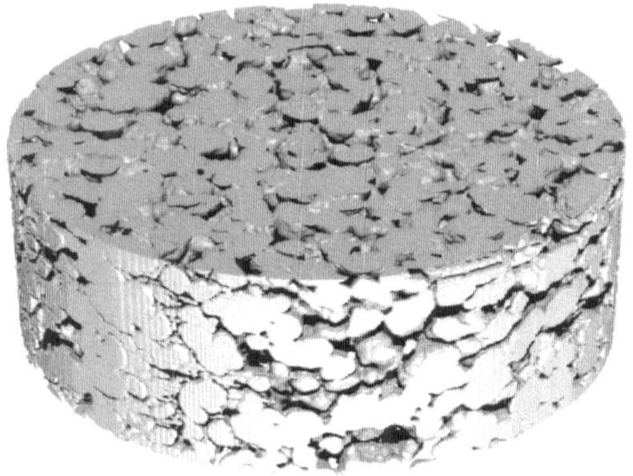

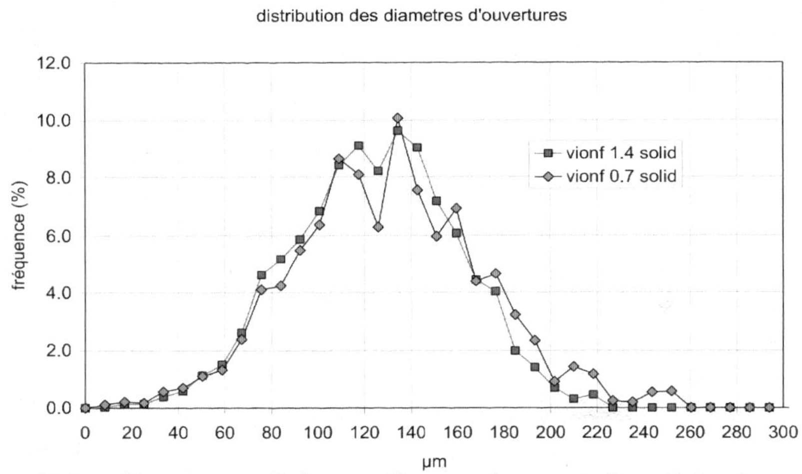

Distribution of the pellet volume of solid phase

Number of pores surrounding a pore

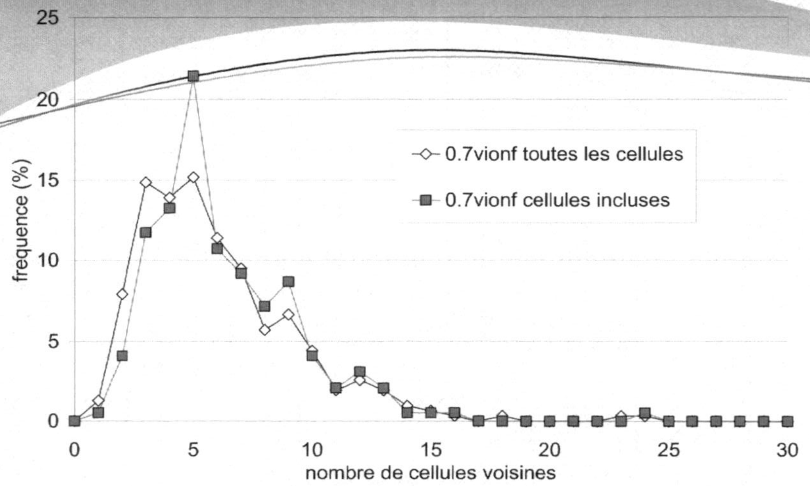

Distribution of number of neighbour pore calculated for each pore

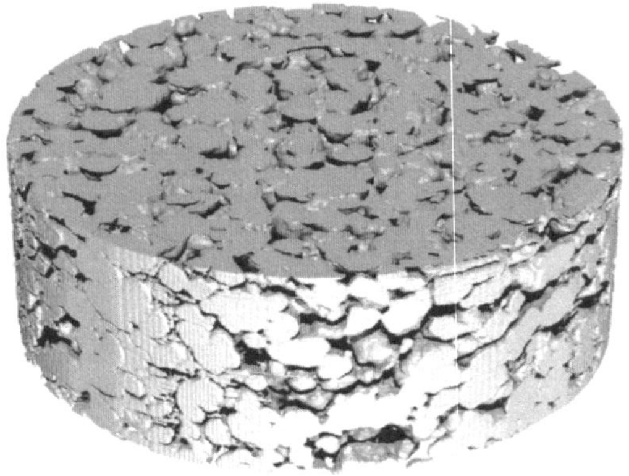

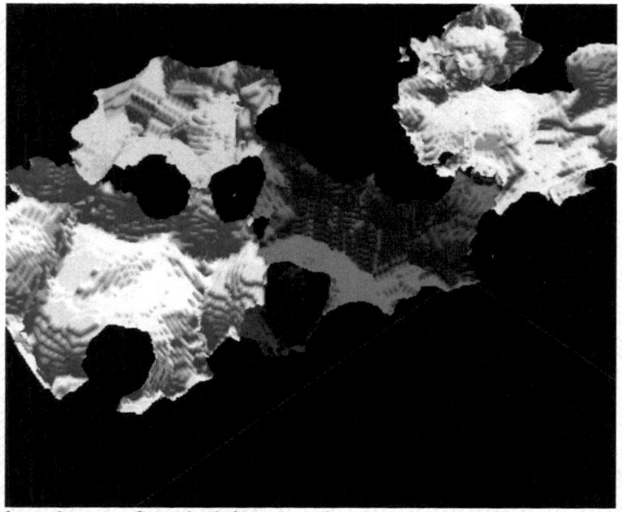

3D Visualization of neighboured pores extracted from liquid phase

Measure of cols

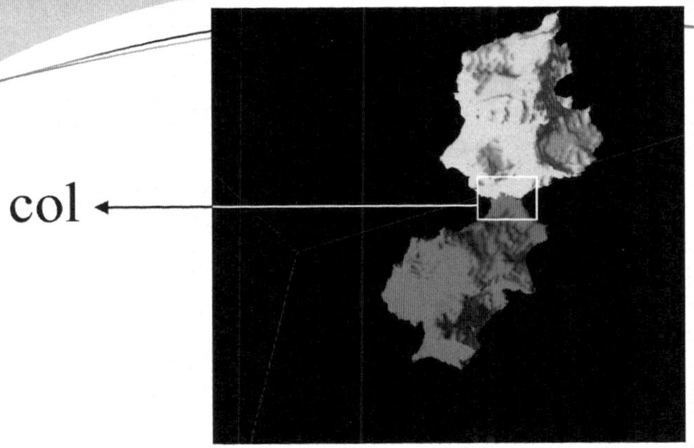

3D Visualization of neighboured pores extracted from liquid phase

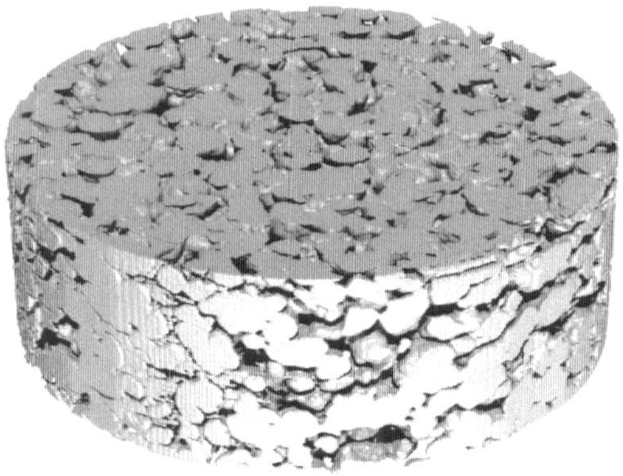

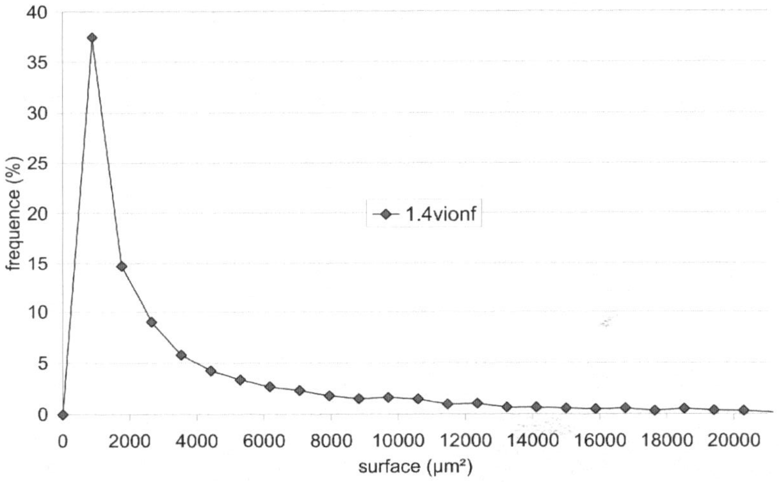

Surface distribution of col of connection between pores

Number of solid element surrounding a single element

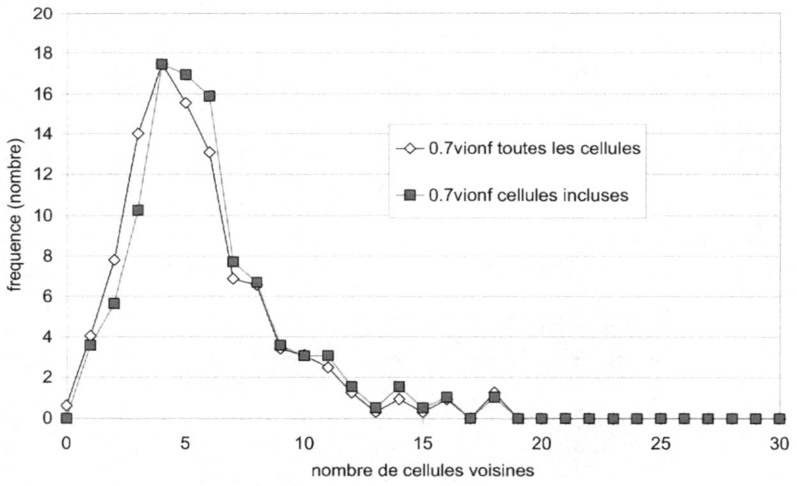

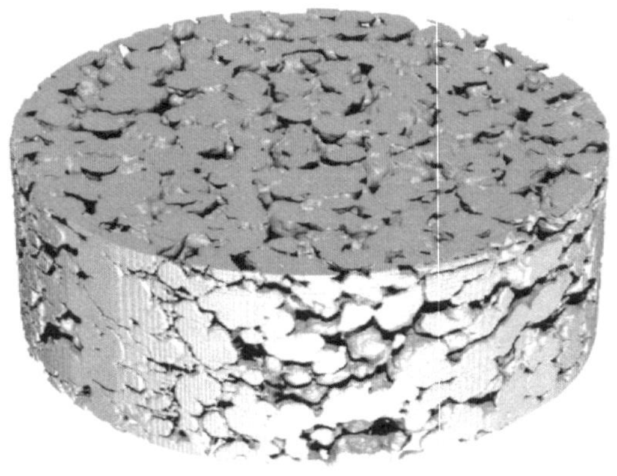

Distribution of number of neighbour solid element calculated for each solid element

Anisotropy

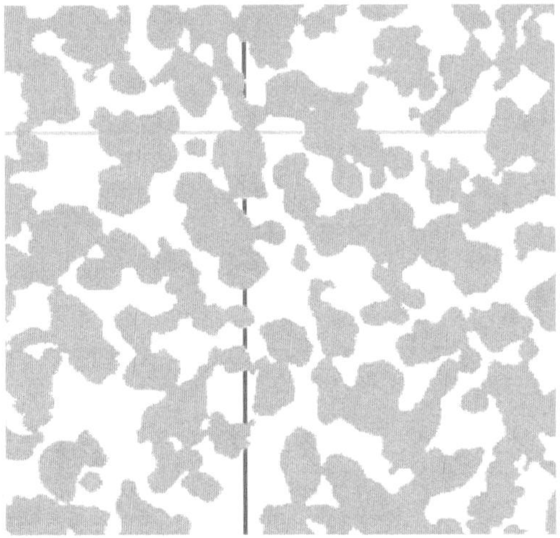

Volume = 0.84mm3
Side = 0.95 mm

This figure represents 1 slice from a cube of Vyon-F having a volume = 0.84 mm3

The red ray intersects 7 times the solid phase when moving in the Z axis (bottom up) which is the direction of the flow

The green ray intersects 6 times the solid phase when moving in the X axis which is the direction transverse to the flow

Agenda

- The challenge : mixing fibrinogen & thrombin
- Using sintered porous polymer as a mixer
- Biological testing to evaluate quality of fibrin
- Assessment of mixer performance with Laser technique
- Testing of the medical device efficiency
- Why does it work? Topology study of the mixer
- Medical & non medical applications
- Conclusion

Manufacturing process evolution

Mix-F device

Artiss RT Mixing Tip

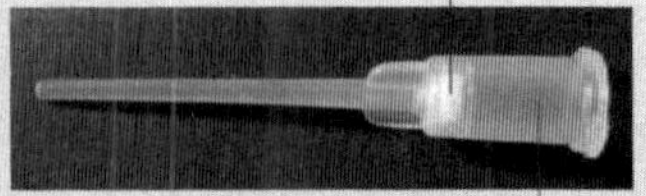

Mix-C device

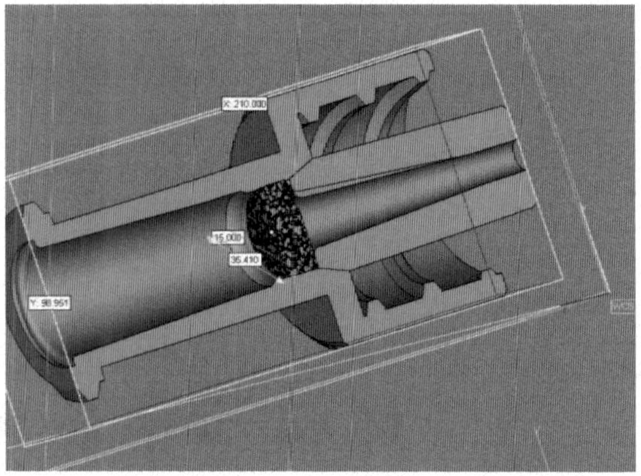

Danette
Fibrin foam
sphère IVIG 600nm
Oil & Diesel
Heterogeneos catalysis
Pet Food
emulsion
Oil & Alcohol
Fibrimex
adhesive agent: beef protein (fibrin)
Batchnr.,
N° lot:
310-05-2805
InLine Mixing process

Conclusion

The porous material fulfills the requirements of European pharmacopoeia, USP, and JPN. The porous material can be sterilized with gamma, beta irradiation, steam, peracetic acid, and ethylene oxide, etc.

The design and configuration of the inline mixing device is optimized for mixing fibrinogen and thrombin for a particular surgical application. The dimensions, shape, porosity, material, conditions of use, however, are not limited to those designated in this application.

Culture of Human Cells in Asymmetric Laminar Flow Systems

Mario M. Alvarez, Leydi Maribel Carrillo-Cocom, David Bulnes-Abundis, Gayathree Murugappan, Marisa Granados-Pastor, Keith Johnson.

Centro de Biotecnología-FEMSA. Tecnológico de Monterrey. Ave. Eugenio Garza Sada 2501. Monterrey, N.L. México, C.P. 64849.

Abstract. In biomedical and tissue engineering applications, such as the expansion of progenitor or endothelial cells, static culture in Petri dishes and bottles is the most common culture strategy. In this contribution, we examine the effect of asymmetric laminar flows of the rate of growth and protein secretion/expression of human stem cells and endothelial cells cultured on Petri dishes. Asymmetric flows within dishes were induced by placing them at different eccentric positions (6.5 and 9,5 cm form the rotational center) in a rotational platform (rotated at 3, 5 and 8 RPM). We observed significant and reproducible growth acceleration of CD34+ umbilical cord-derived human hematopoietic progenitor cells when exposed to rotational flow as compared to stationary conditions. Growth dynamics under 3 or 5 RPM agitation was compared against that observed under typical stationary conditions. Cells cultured at 3 or 5 RPM exhibited (a) the absence of a latency phase, (b) an increase in final cell concentrations by 54%-58.5%, and (c) reduced doubling time in their exponential phase by 12-16% in comparison to stationary culture. Cells grown under rotational agitation were confirmed to remain CD34+ by PCR. These results document a significant positive effect of exposure to laminar flow fields on the growth of human hematopoietic progenitor cells. In experiments with endothelial cells, significant differences in both, secretion level and glycosylation patterns (extent and completeness of glycosylation) were observed as a result of relatively small changes in rotational speed. These results support the hypothesis that differences in hydrodynamic conditions may induce important changes in the secretion and assembly of glycoproteins in mammalian cell cultures.

Keywords: Mammalian cell culture, mixing, Petri, bioreactors, Biotechnology, stem cells, endothelial.

1. INTRODUCTION

Human cell culture is already a routine operation in Biotechnology laboratories around the world. For tissue engineering applications, mammalian and particularly human cell culture is a core activity. Most recently, the highly promising medical potential of stem cell technology has made more evident the need for perfecting stem cell culture techniques. Also, for more basic applications, such as the understanding of the effects of hydrodynamic conditions on the biochemical behaviour of the endothelial cells that cover our veins and arteries, mammalian cell culture is required.

In this contribution, we examine the effect of mixing at low speed (in the laminar regime) in the rate of growth and the biochemical expression of different types of anchorage dependant human cells. In particular, two types of human cells were selected as experimental models. Human stem cells obtained from bone marrow were used as a second experimental model. A faster growth of human stem cells is key for the commercial autologous stem cell transplantation in humans to be technically feasible. We demonstrate that laminar agitation has a significant effect on the growth rate of human stem cells cultured in Petri dishes. Cells cultured in this way do no exhibit morphological or biochemical signs of differentiation. In addition, endothelial cells are used as a third experimental model. Our veins and arteries are covered with endothelial cells. Due to their natural role as secretors of proteins that control coagulation and other crucial vascular processes, endothelial cells have evolved to be highly responsive to hydrodynamic conditions. Here we demonstrate that, not only the level of secretion of an important coagulation protein expressed by endothelial cells (the von Willebrand Factor or vWF) but also its glycosylation patterns are highly dependent on agitation conditions.

2. MATERIALS AND METHODS

Experimental set up

Cells were cultured in 6 mm diameter Petri dishes (from Corning®, USA). Culture vessels were agitated in a rotating circular platform (Figure 1). Four different rotational speeds were tested (0, 3, 5, and 8 rpm). Also, culture vessels were placed at different radial distances with respect to the rotational axis.

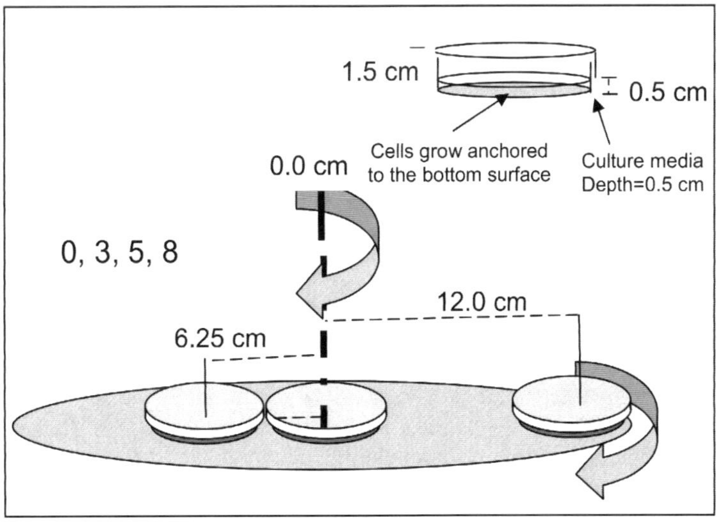

Figure 1. Experimental set up. Petri dishes were agitated at 0, 3, 5, or 8 RPM on a rotating circular platform

Cell culture techniques

Two different cell types were used in our experiments. Human hematopoietic progenitor cells were isolated from umbilical cord blood by diluting the blood 1:1 with PBS-2mM EDTA buffer (GIBCO, United States) according to standard protocols (Murugappan *et al.*, 2009). Cells were seeded in 75 cm^2 culture flasks (Corning) in 15 ml DMEM-F12 supplemented culture media and incubated for 24 hrs at 37°C and 95% relative humidity in a 5% CO_2 and 95% air atmosphere. All cells used in the reported experiments correspond to the third subculture from the cells originally isolated.

Human Umbilical Vein Endothelial Cells (HUVEC) were isolated from human umbilical cords following protocols detailed elsewhere (Rodríguez *et al.*, 2002; Rivas-Arreola, 2005; Morga-Ramírez, 2007). Cells were seeded in 25 cm^2 culture bottles (Corning) in 4 ml of supplemented M199 culture media (previously described) and incubated at 37°C and 95% relative humidity in a 7% CO_2 and 93% air atmosphere. All endothelial cells used in the reported experiments correspond to the third subculture from the cells originally isolated.

Protein expression experiments

Released vWF antigen was measured in samples taken from the culture supernatant of the Petri dishes using a sandwich immune-enzymatic assay detailed by Morga-Ramírez (2007).

In order to investigate the effect of low Re rotational agitation on the glycosylation patterns of vWF, we measured the total concentration of trimannose and terminal sialic acid residues on vWF released by confluent cultures of endothelial cells agitated at different rotational speeds. Mannose and sialic acid content was calculated by an enzyme lynked lectin assay (ELLA) slightly modified from that reported by Collados *et al.* (1999).

CFD experiments

Computational fluid dynamics (CFD) was used to characterize flow behavior within Petri dishes rotating at different velocities. A solid model of the initial fluid volume of the Petri dish was made using Synergy, from Weber Systems Inc., Menomonee Falls, WI, and imported into the solver environment used in this work (ACUSOLVE CFD) using the acuConsole program also from ACUSIM Software Inc. The mesh was generated using the acuMeshSim program, which also facilitated the generation of very refined boundary layer elements from the dish surface. The resultant model consisted of over 755,000 tetrahedra elements, and 137,000 nodes. The fluid was modelled primarily as a Newtonian fluid with the nominal properties of water.

3. RESULTS AND DISCUSSION

Exposure to rotational laminar velocity fields of human stem cells cultured in 6mm Petri dishes at 3 RPM and 5 RPM caused a significant decrease in cell doubling time in comparison to stationary culture (0 RPM). Cells cultured at radial distances of 6.25 cm and 12.5 cm at 3 RPM exhibited a 200% shorter latency phase and yielded 46% and 62% higher cell concentrations, respectively, at the end of the 10-day culture period (Figure 2). For the duration of the experimental period, overall cell viability was maintained at 0.7%, 2.6% and 1.89% higher by agitation at 3 RPM at radial distances of 0 cm, 6.25 cm and 12.5 cm, respectively. Doubling time was calculated using linear regression analysis of the exponential growth phase. Rotation at radial distances of 6.25 cm and 12.5 cm reduced doubling time in the exponential phase by 19.05% and 18.19%, respectively. Rotation at 0 cm radial distance decreased doubling time by 10.76% in comparison to stationary culture (0 RPM).

At 5 RPM agitation, cells observed a 100% shorter latency period and yielded 57% and 60% higher cell concentrations at radial distances of 6.25 and 12.5 cm, respectively. Rotation at radial distances of 6.25 cm and 12.5 cm reduced doubling time in the exponential phase by 33% and 25%, respectively, in comparison to stationary culture. For the experimental period, cell viability remained on average 1.53%, 4.15% and 4.40% higher at radial distances of 0 cm, 6.25 cm and 12.5 cm, respectively. For rotation at both 3 RPM and 5 RPM, placement of Petri dishes at a radial distance of 0 cm caused an increase in doubling time of 10.73% and 10.82%, respectively.

No significant morphological changes were observed after exposing cells to a rotational laminar velocity field at 3 (see Figure 2c and 2d) or 5 RPM at radial distances of 0 cm, 6.25 cm and 12.5 cm for 10 days.

Cells grew in monolayer, and were easily detachable by typical trypsin treatment. However, cells grown in suspension in a 30 ml stirred tank exhibited agglomeration. Upon completion of the 10-day rotational experiment, RNA was extracted from cells at each radial distance (0, 6.25 and 9.5 cm) as well as the control cells in stationary culture (0 RPM). RT-PCR products in agarose gel electrophoresis results (not shown) revealed that all RNA profiles remained positive for CD34 and β-actin following agitation at 3 RPM.

In our experiments with endothelial cells, we also observed significant differences between agitated and non agitated cases. Rotational speeds in the low Reynolds regime (low Re) have a significant effect on the levels of secretion and the glycosylation patterns of the von Willebrand factor expressed by HUVEC cultures. Low rotational velocities (3 RPM; shear stresses $\pi_{xy} \leq 55$ dynes/cm^2; Re ≤ 30) promote higher secretion of vWF than static cultures. In the range of Re analyzed here, as shear stress (π_{xy}) increases, vWF increases (as reported before by Galbusera *et al.*, 1997).

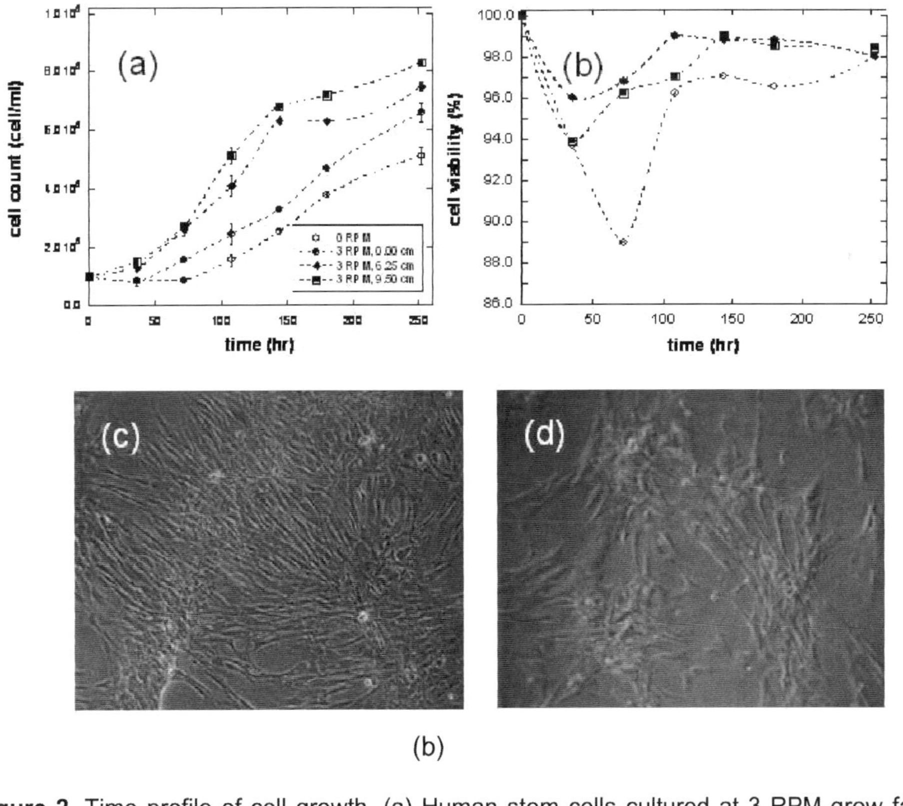

(b)

Figure 2. Time profile of cell growth. (a) Human stem cells cultured at 3 RPM grow faster than at static conditions; (b) viabilities are also higher at agitated than static conditions. The morphology of cell cultures observed at static conditions (c) is similar to that observed in rotational culture systems (d).

However, not intuitively, rotational flows of 5 RPM (5 RPM; $\pi_{xy} \leq 37$ dynes/cm^2; Re $\leq$ 50) induce a significantly higher extent of glycosylation (measured as a content of trimannose core structures per vWF unit) with respect to static experiments (Re=0) or experiments at 8 RPM (8 RPM; Re>30; $\pi_{xy\ max}$ =55 dynes/cm^2).
Evaluation of glycosylation completeness, based on the sialic acid content of vWF secreted at different rotational speeds, indicate that small variations in Re values (or shear stress) have significant effects. While the content of sialic acid residues is superior in experiments at low and medium rotational speeds (5 RPM: Re<50; $\pi_{xy\ max}$ < 37 dynes/cm^2) with respect to static conditions, in experiments at higher rotational speeds (8 RPM; Re>80; $\pi_{xy\ max}$ = 55 dynes/cm^2) glycosylation completeness was poorer than in static conditions (Figure 3).

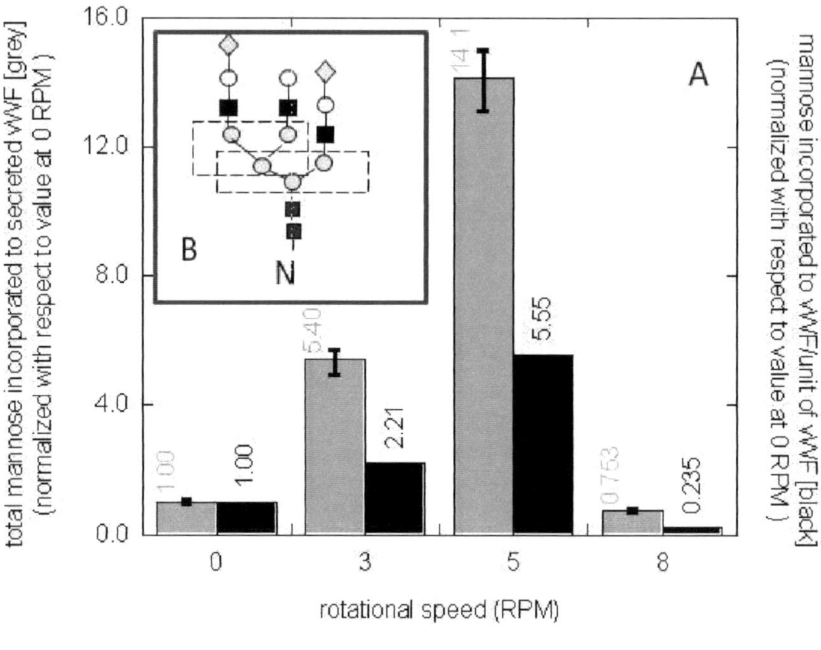

Figure 3. Glycosylation extent at different rotational conditions, as determined by a Concanavalin A binding assay to trimannose structures. A) Glycosilation extent normalized with respect to static conditions (black bars), and normalized with respect to static conditions and vWF secretion (gray bars). Each bar represents results from 3 to 4 replicas. B) Position of trimannose core structures in glycans.

CFD calculations revealed that the flow field and the distribution of shear stresses within the Petri dishes are highly complex (see for example Figure 4). Flow is fully 3D and asymmetrical, which favors mixing and adequate mass transfer. However, the flow field is also highly heterogeneous. In some regions and at some conditions, high local speeds and shear rates can be observed. This in turn, might make it difficult for anchorage dependent cells to adhere.

In particular, a high shear rate zone at the center of Petri dishes rotating at 8 RPM is consistent with the poor glycosylation performance observed in endothelial cells growing in that condition.

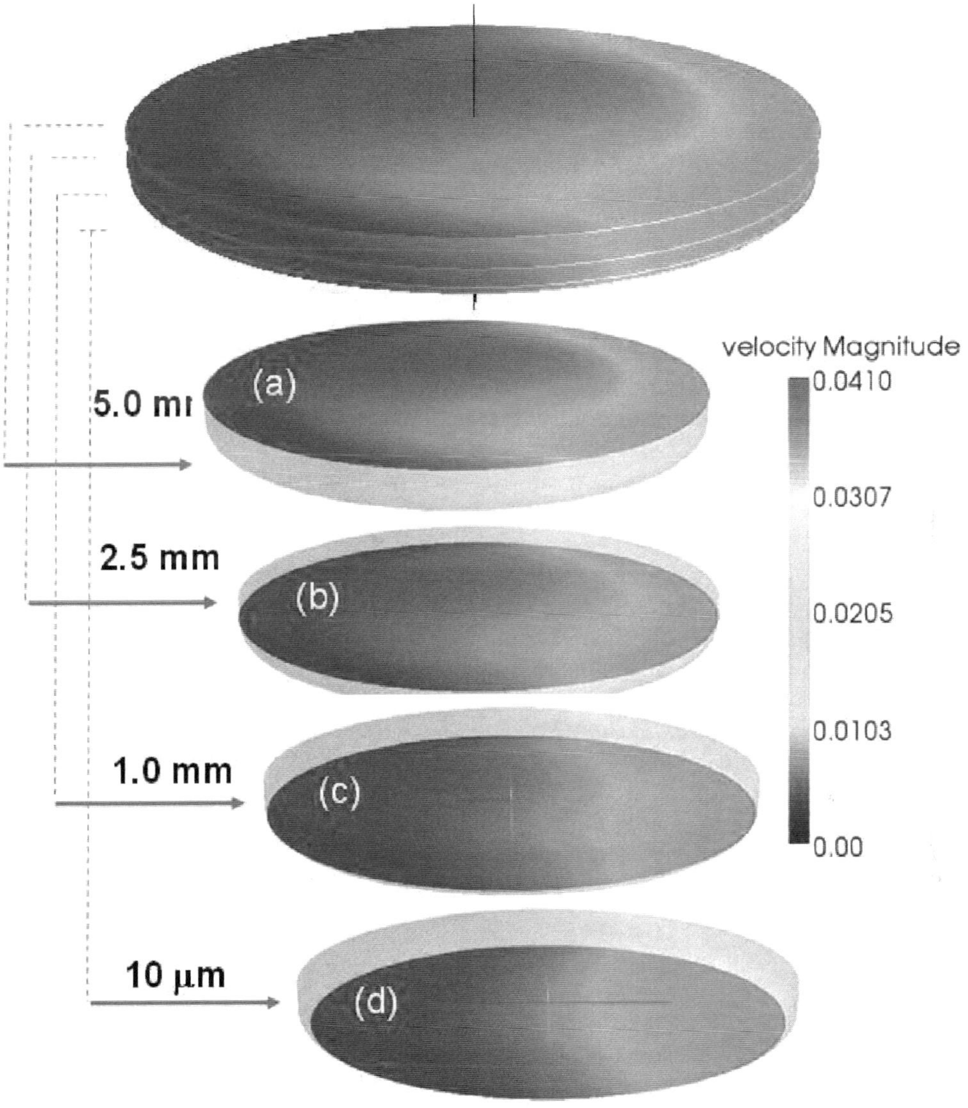

Figure 4. For dishes eccentrically located within the rotating platform, flow structure is asymmetric and 3D as shown by CFD velocity fields at different planes of a Petri dish rotated at 3RPM and located at 6.25 cm from the rotational center: (a) 4.9 mm from the bottom surface (slightly bellow the liquid surface); (b) 2.5 mm from the bottom surface; (c) 1.0 mm from the bottom surface; and (d) 10 μm from the bottom surface (slightly above the cell monolayer).

4. CONCLUSIONS

In this contribution we document significant effects of mixing in the rate of growth and the biochemical expression of anchorage dependent cells. The specific growth rate of BM-MSCs (rat stem cells) cultured in 96-wells culture plates, and BM-HSC (human stem cells derived from bone marrow) cultured at 6 mm Petri dishes is positively affected by rotational agitation at low speed (3 RPM and 3 or 5 RPM respectively). In the case of endothelial cells, the secretion level of the vWF progressively increased significantly when rotated at 3, 5 and 8 RPM. The glycosylation patterns of this glycoprotein, specifically the trimannose and sialic acid content, are also significantly affected by rotational agitation from 3 to 8 RPM.

REFERENCES

Carrillo-Cocoom, L.M. (2006). Evaluación del efecto del estrés mecánico en la cinética y en la expresión proteica de células madre mesenquimales de médula ósea de rata. *Master Degree Thesis.* Tecnológico de Monterrey. México. 109 pp.

Collados, M.T., J. Sandoval, S. López , F.A. Massó, A. Páez, J.R. Borbolla and L.F. Montaño (1999). Characterization of von Willebrand factor in primary pulmonary hypertension. *Heart Vessels* 14:246-252.

Galbusera, M., C. Zoja, R. Donadelli, S. Paris, M. Morigi, A. Benigni, M. Figliuzzi, G. Remuzzi, A. Remuzzi. (1997). Fluid shear stress modulates von Willebrand factor release from human vascular endothelium. *Blood* 90(4): 1558-1564.

Morga-Ramírez M. (2007). Efecto del Estrés mecánico o de Concentraciones Elevadas de Homocisteina sobre los Patrones de Glicosilación del Factor von Willebrand en Células Endoteliales *in vitro. Master Degree Thesis.* Tecnológico de Monterrey. México. 93 pp.

Rivas Arreola, J.M. (2005). Patrones de colonización de células endoteliales y células madre en discos rotatorios. *Master Degree Thesis.* Tecnológico de Monterrey. México. 48 pp.

Rodríguez, E., R. López, A. Páez, F. Massó, and L.F. Montaño. (2002). 17β-estradiol inhibits the adhesion of leukocytes in TNF stimulated human endothelial cells by blocking IL-8 and MCP-1 secretion, but not its transcription. *Life Sciences* 71(18): 2181-2193.

Comparison of Computed and Measured Wall Shear Stress in An Orbiting Cell Culture Dish

Jonathan Thomas, Dept. of Chemical Engineering, University of Louisville, Louisville, KY

M. Keith Sharp, Dept. of Mechanical Engineering, University of Louisville, Louisville, KY

R. Eric Berson, Dept. of Chemical Engineering, University of Louisville, Louisville, KY

It is well documented that physiological and morphological properties of anchored cells are influenced by fluid shear stress. Common orbital shakers provide a means of mixing the fluid containing nutrients while simultaneously applying shear stress to cells for tens to hundreds of cases by loading the shaker with multiple dishes. However, the complex flow in orbiting dishes is generally not amenable to analytical solution for resolving shear created by the fluid motion. The only existing quantification of shear in this flow is an equation that estimates a constant scalar value of shear for the entire surface of the dish. In practice, wall shear stress (WSS) will be oscillatory rather than steady due to the travelling waveform and will vary across the surface of the dish at any instant in time. Here we present a computational model that for the first time provides complete spatial and temporal resolution of WSS over the bottom surface of a dish throughout the orbital cycle. The model is reasonably well validated by both the analytical solution (resultant WSS magnitudes that are within 0.99 ± 0.42 dyne/cm2) and comparison to tangential WSS magnitudes obtained using one-dimensional optical velocimetry at discreet locations on the bottom of an orbiting dish. The experimental maximum WSS varied by about 12% from the computational model at both locations that were measured.

Neglecting surface tension, flow in the dish is influenced by Reynolds, Froude and Stokes numbers and a slope parameter [ratio of steady-state acceleration-induced free surface slope and static fluid aspect ratio (fluid height to dish radius)]. Conditions on a shaker table are typically low Reynolds and low slope. The existing analytical result is an extension of Stokes 2nd problem to orbital plate motion for low Froude and high Stokes, and is valid for flow not too close to the vertical walls of the dish, which was confirmed by CFD results. These results help explain why cellular responses depend on location in the dish.

See more of this Session: Novel Experimental Methods in Mixing Research
See more of this Group/Topical: North American Mixing Forum

Heat Exchange in the Rotated Arc Mixer: Experiment, Theory, and Plant Trials

Guy Metcalfe, Materials Science and Engineering, CSIRO, Melbourne, Australia

Daniel Lester, Materials Science and Engineering, CSIRO, Melbourne, Australia

Murray Rudman, Mathematical and Information Science, CSIRO, Melbourne, Australia

The Rotated Arc Mixer (RAM) is a continuous mixer for highly viscous Newtonian or shear sensitive fluids. By using programmed boundary motions to generate chaotic advection instead of internal obstructions means the RAM's pressure drop is the same as an open pipe. Previous work demonstrated the RAM's efficacy as a mixer [1]. By adding a thermal jacket the RAM becomes a heat exchanger for highly viscous products. However, the change from solving the conservative kinematic equation to solving the dissipative advection-diffusion equation changes the basic nature of the optimization problem for designing the boundary motions to maximize heat transfer rates and global temperature equilibration. The thermal optimization problem for chaotic stirred flows typically results in a fractal distribution of the performance enhancement over the parameter space of the flow, and we have developed new methods to resolve this space and solve the optimization problem [2,3]. Lab experiments and theory achieve a nearly 80% reduction in total energy input over an equivalent pipe required to thermalize viscous Newtonian fluid. We will also present results from pilot trials done with a food processing company [4].

[1] Metcalfe et al. 2006. AIChE J., 52(1), 9--28, doi:10.1002/aic.10640.

[2] Lester et al. 2007. J. Comp. Phys., 227, 3032--3057, doi:10.1016/j.jcp.2007.10.015.

[3] Lester, Rudman & Metcalfe. 2009. Int. J. Heat Mass Trans., 52, 655--664, doi:10.1016/j.ijheatmasstransfer.2008.06.039.

[4] Metcalfe & Lester. 2009. J. Food Eng., in press, doi:10.1016/j.jfoodeng.2009.04.032

See more of this Session: Novel Experimental Methods in Mixing Research
See more of this Group/Topical: North American Mixing Forum

Energy Savings and Improved Mixing Performance with Modified Impeller Technology in Paper Pulp Mixing

Marc Moseley, Philadelphia Mixing Solutions Ltd, Palmyra, PA

Wojtek Wyczalkowski, R&D, Philadelphia Mixing Solutions Ltd, Palmyra, PA

Pulp and paper plants use stock chests with partial mixing and recirculation loops to reduce the magnitude of consistency fluctuations in pulp production. Minimizing these fluctuations provides smooth, uniform surfaces; improves final sheet properties; and, most importantly, offers substantial, measurable cost savings.

At present, most stock chests are only partially mixed with side-entry mixers. Maximum consistency is limited to less than 6.5% due to mixing limitations. New data shows that by mixing the entire volume of the stock chest, consistency fluctuations can be minimized. This was not feasible in the past due to lack of suitable mixing technology. However, with the proper application of modified impeller technology combined with zonal mixing, the entire chest can be uniformly mixed using only 50% of the power required by traditional side-entry mixers. This technology will allow for uniform mixing of paper pulp up to 14% consistency. Independent estimates of savings due to improved sheet profiles, increases in production efficiencies and energy savings vary, but can exceed $100,000.00 per month (USD).

See more of this Session: Novel Experimental Methods in Mixing Research
See more of this Group/Topical: North American Mixing Forum

Using Computational Fluid Dynamics with Experimental Validation to Predict Liquid-Solid Mixing in Vessels Equipped with Pulsed-Jet Mixers

Brigette Rosendall, Advanced Simulation and Analysis, Bechtel National, Inc., Thousand Oaks, CA

Milorad Dordevich, Advanced Simulation and Analysis, Bechtel National, Inc., San Francisco, CA

YoChan Kim, Advanced Simulation and Analysis, Bechtel National, Inc., Frederick, MD

Chris Kennedy, Advanced Simulation and Analysis, Bechtel National, Inc., San Francisco, CA

The Columbia River in Washington State is threatened by the radioactive legacy of the cold war. Two hundred thousand cubic meters (fifty-three million US gallons) of radioactive waste is stored in 177 underground tanks (60% of the nation's radioactive waste). This waste is a product of 50 years of plutonium production for national defense. Bechtel National, Inc (BNI) has been commissioned by the U.S. Department of Energy to design and build a vast complex of waste treatment facilities to convert this waste into stable glass using a proven vitrification process.

The waste in these underground storage tanks is a combination of sludge, slurry, and liquid. Part of the engineering challenge for during processing is to verify that mixing systems in the plant vessels keep all solids in suspension. For various reasons, the mixing system chosen is one based on jet mixers. Specifically, these vessels use compressed air driven pumps to power what are known as pulsed-jet mixers (PJMs).

Computational Fluid Dynamics has been applied to the study and design of Pulse Jet Mixing systems at the Hanford Site Waste Treatment Plant since 2001. Several experimental studies have been commissioned by the Waste Treatment Plant project to understand the ability of PJMs to keep solids in suspension and determine levels of slurry blending. This work will present CFD models of these experimental platforms. These models are intended to benchmark the CFD for use in design analysis.

Eulerian-granular CFD models of a 15-inch and a 70-inch diameter test vessel will be presented. The tests were performed at just-suspended velocities and concentration measurements were recorded at various levels in the vessel. Time history comparisons between the experiments and CFD results will be presented along with other parameters intended to show how CFD can be used to predict suspension and blending.

See more of this Session: Novel Experimental Methods in Mixing Research
See more of this Group/Topical: North American Mixing Forum

Mixing Performance of a Novel, Continuous Confined Impinging Jets Reactor Using Competitive Reactions

Han Zheng, Chemical, Biological and Pharmaceutical Engineering, New Jersey Institute of Technology, Newark, NJ

Giuseppe Di Benedetto, Otto H. York Department of Chemical, Biological and Pharmaceutical Engineering, New Jersey Institute of Technology, Newark, NJ

Piero Armenante, Chemical, Biological and Pharmaceutical Engineering, New Jersey Institute of Technology, Newark, NJ

In this work, a novel continuous flow apparatus featuring the impingement of fluid jet streams in the presence of ultrasonic energy provided by an ultrasonic probe was tested using competitive reactions in order to determine its mixing effectiveness. The ultrasonic energy enhances micromixing of the fluid jet streams, which results in an overall mixing effectiveness improvements in different physical and chemical processes. Such an apparatus has been shown by our group to provide improvements in crystallization and precipitation processes, as well as fast chemical reactions and selectivity improvement.

In the competitive reaction system used here (third Bourne reaction), one stream containing sodium hydroxide was continuously fed to the first the impinging jet, while an aqueous solution of hydrochloric acid and ethyl chloroacetate was fed to the second jet. At the end of each experiment, the concentration of ethanol (obtained from the conversion of the slower ethyl chloroacetate reaction) was experimentally determined by gas chromatography (GC) to determine the mixing efficiency: higher ethanol concentrations implied poorer mixing. In all experiments, the volumetric flow rate of the limiting solution, i.e., the sodium hydroxide solution, was always kept the same, while the flow rate of the other solution was changed while also changing, at the same time, their concentrations in order to ensure that the reactants were in the appropriate stoichiometric ratios. The sonication power was also varied.

It was experimentally found here that improved mixing was achieved by increasing volumetric flow rate with or without sonication. If sonication was applied, mixing also improved significantly and it was further improved when a higher sonication power was applied. This trend was especially evident at lower flow rates. At high flow rate, sonication played a smaller role. A quantitative comparison of the results based on the use of the Damkohler number is provided.

See more of this Session: Novel Experimental Methods in Mixing Research
See more of this Group/Topical: North American Mixing Forum

Direct Flow Simulations of Thixotropic Liquids in Agitated Tanks

Jos Derksen, Chemical & Materials Engineering, University of Alberta, Edmonton, AB, Canada

Yield-stress liquids often show time-dependent behaviour (thixotropy): the network providing the yield stress in the liquid needs time to build up after being destroyed by liquid deformation. In numerical simulations of thixotropic liquids the network's integrity can be accounted for by solving a scalar transport equation in a network structure parameter, and relating that structure parameter to the apparent viscosity. We apply this method for simulating the laminar and transitional flow in a baffled mixing tank equipped with a Rushton turbine with emphasis on the start-up of the flow from a stagnant situation, and on comparing liquids with time dependent and time independent rheology.

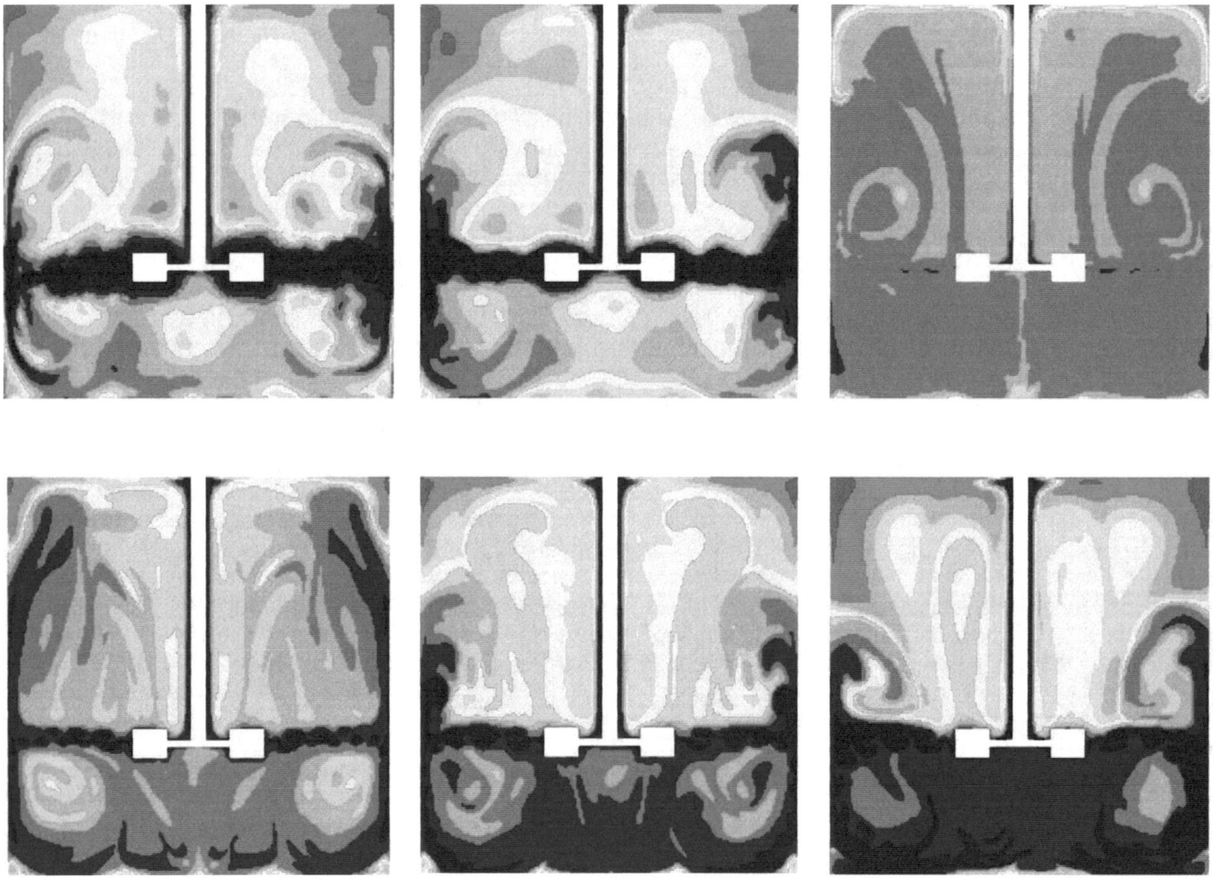

See more of this Session: Novel Computational Methods in Mixing Research and Design
See more of this Group/Topical: North American Mixing Forum

Simulation of Flow Field and Circulation Time Phenomena in a Batch Rotor-Stator Mixer

Meng Yang, Chemical & Biomolecular Engineering, University of Maryland, College Park, MD

Richard V. Calabrese, Chemical & Biomolecular Engineering, University of Maryland, College Park, MD

Batch rotor-stator mixers are widely employed on the bench scale in the chemical and pharmaceutical process industries to demonstrate a variety of emulsion and dispersion processes, that must be scaled to geometrically or dynamically (continuous rather than batch) different equipment on the pilot scale. When the process fluid dynamics plays a significant role, a reasonable approach is to simulate both the bench and pilot scale flow field and deformation rate phenomena and match these on both scales to extrapolate bench scale data with respect to the number of mill head passes and shear rate history.

To this end, we are simulating the flow field in a Silverson L4R batch rotor-stator mixer with the milling head located off-axis in a 2 liter vessel. Since water is the working fluid, the flow is turbulent. We use a RANS formalism with a k - epsilon turbulence model and wall functions. Due to the importance of the rotor-stator interaction, a sliding mesh CFD technique must be employed. While time periodic convergence is relatively rapid close to the mill head, the pumping capacity is low. As a result, it is not possible with current computational resources to compute until a fully converged solution is reached throughout the entire tank. This is of little consequence if the goal is to understand rotor-stator interactions. However, it is a major barrier to allowing one way coupled particle tracking throughout the vessel, to determine the shear history and circulation time distribution of tracked particles.

At a moderate distance from the mixing head the effect of rotor-stator interactions is not felt, so the flow appears steady in time. A time periodic region, centered on the mill head can be defined by interrogating a sliding mesh solution for the entire tank. Beyond this region, the flow field is well described by an MRF simulation technique. As a result, the following hybrid technique can be used to significantly reduce the computational expense of particle tracking simulations. Based on mixing head geometry/placement, vessel size/geometry, rotor speed and physical properties, the largest possible time periodic region can be defined. We then begin with an accurate, fully converged MRF solution, which always holds beyond the time periodic (outer) region and provides boundary conditions for a sliding mesh solution within the time periodic (inner) region. The sliding mesh algorithm is then begun for the inner region. This inner region is relatively small in volume, so convergence is rapid. The inner and outer solutions are then super imposed to provide a seamless time dependent velocity field for particle tracking trials. A single, repeatable period (in this case 90 degrees due to 4 rotor blades) of this simulation can be continually repeated as input to a fast particle tracking algorithm. Our previously reported computationally efficient tracking algorithm, which accounts for all of the forces acting on particles, can then calculate the trajectories of thousands of particles for numerous impeller revolutions to yield statistics for circulation time distribution and shear rate history.

We will demonstrate the hybrid velocity field CFD technique and particle tracking algorithm, and report the results of our simulations for several different cases. We will discuss our future plans in the context of our progress to date and what can realistically be accomplished.

See more of this Session: Novel Computational Methods in Mixing Research and Design
See more of this Group/Topical: North American Mixing Forum

Simplified Model-Based Design and Shape Optimization for T-Shaped Microreactors with Secondary Flow

Lin Wang, Department of Chemical Engineering, Kyoto University, Kyoto, Japan

Osamu Tonomura, Department of Chemical Engineering, Kyoto University, Kyoto, Japan

Manabu Kano, Department of Chemical Engineering, Kyoto University, Kyoto, Japan

Shinji Hasebe, Department of Chemical Engineering, Kyoto University, Kyoto, Japan

T-shaped microreactors, which use two horizontal arms as the inlet channels and one vertical arm as the mixing channel, have attracted in the academic and industrial field. Using T-shaped microreactors in secondary flow regime, a complete reaction with a high product yield can be realized on a short mixing time. In the design of T-shaped microreactors to achieve desired product yield, the fluid velocity and the size and geometry of mixing channel are essential factors, but a systematic design method on these essential factors has not been established. Moreover, when parameter studies or optimization runs need to performed in the microreactors, the repeating computational burden related to the normal Computational Fluid Dynamics (CFD) model is usually much too heavy. Therefore, a simplified model is indispensable to describe the mixing, diffusion and reaction phenomena in the T-shaped microreactors.

In this research, a simplified lamellar model is developed to efficiently design T-shaped microreactors with secondary flow. In the model, the same thicknesses of striations are produced initially by mixing without reaction or diffusion, and then reaction and diffusion are allowed to occur in the mixing channel. The concentration profiles under laminar flow condition in the flow direction are estimated by a simplified lamellar model based on the two-dimensional convection- diffusion -reaction equations. The two dimensions are the flow direction along the mixing channel (x direction) and the same direction as the flow path along the inlet channel (y direction). At the first step of building the lamellar model, the fluid velocity and the striation thickness are used to predict the product yield. Then, a pseudo-three-dimensional model, which combines the x-y lamellar model and the x-z lamellar model (z is the vertical direction of x-y surface), is constructed. It can complement the concentration distribution in the y-z direction of cross section at different location along the mixing channel. Therefore, comparing to the first x-y dimensional lamellar model, the predictive accuracy of concentration profiles is improved by the pseudo-three-dimensional model.

Using the developed pseudo-three-dimensional lamellar model, the cross section size can be optimized to get the desired product yield. To investigate the effectiveness of the optimization results, a series-parallel reaction system in T-shaped microreactors is used to compare the three-dimensional CFD simulation results and the proposed model. In the comparison, the rate constant of the reaction and the flow velocity distribution are discussed on the effectiveness of the lamellar model. Though the discussion results, a design guideline for shape design using the simplified model is mentioned.

See more of this Session: Novel Computational Methods in Mixing Research and Design
See more of this Group/Topical: North American Mixing Forum

DNS of a Turbulent Stirred Vessel with Baffles and a Rushton Turbine

Jurriaan J.J. Gillissen, Dept of Multi-Scale Physics, Delft University of Technology, Delft, Netherlands

Harry E.A. Van den Akker, Dept. of Multi-Scale Physics, Delft University of Technology, Delft, Netherlands

We report on a Direct Numerical Simulation (DNS) for a stirred tank provided with four baffles and a six-bladed Rushton turbine, operated in the turbulent regime (Re = 7,300). The incompressible and isothermal Navier-Stokes equations are discretized using a Lattice Boltzmann method and the no-slip boundary condition is enforced using an adaptive force field technique. To resolve all turbulent scales during 20 turbine revolutions, a total of 4 billion grid-points and 10e5 time-steps are computed. The job was performed during one month on 128 CPU cores.

In the talk, we focus on a comparison between LES and DNS. We point out where LES fails and suggest ways for improving the sub-grid modeling. Also we show visualizations of the vortical structures in the wake of a single turbine blade.

See more of this Session: Novel Computational Methods in Mixing Research and Design
See more of this Group/Topical: North American Mixing Forum

Correlation for Yield of Competitive Reactions with Turbulent Mixing

Gary K. Patterson, Department of Chemical Engineering, Missouri University of Science and Technology, Rolla, MO

A correlation for yields of products of competitive reactions with turbulent mixing resulted in a plot of [yield/perfect yield] versus a mixing Damkoehler number, the ratio of mixing time to reaction time, written as follows: DaM = TauM/TauR, where TauM = 2(L^2/Eps)^1/3 and TauR = 1/(k1 k2 CAf CBres)^1/2. Such competitive reactions are represented by A + B --> R; R + B --> S and A + B --> R; C + B --> S. The scale of mixing L is approximated in a stirred vessel by Df = (4Qf/Pi Ubt)^1/2 and the rate of turbulence energy dissipation rate in a stirred vessel Eps is given by Ni NP D^2 N^3/[(T/D)^3(Pi/4)]. [Symbols are defined as follows: k1 and k2 are rate constants for reactions 1 and 2, CAf and CBres are molar concentrations of the feed A and the resident B, Df is the diameter of the feed jet as stretched or compressed by the flow, Qf is the volumetric feed rate, Ubt is the blade tip speed of an impeller, Ni is the ratio (feed point/average) , NP is the impeller power number, N is the impeller rotation rate, T is the vessel diameter, and D is the impeller diameter.] The same approach also works in many turbulent pipe reactors where the geometry of the feed injection is well known.

A recent more detailed dimensional analysis has shown the need to include the effect of the ratio of reaction rate constants, k2/k1, in the correlation, even though the ratio of yield to perfect yield removes most of the divergence. The new correlation effectively separates experimental data for different k2/k1-ratios. Initial work to extend the correlation method to solid-liquid reactions with turbulent mixing will also be presented.

See more of this Session: Novel Computational Methods in Mixing Research and Design
See more of this Group/Topical: North American Mixing Forum

Simplification and Application of An Interfacial Area Concentration Balance by Order Arguments

Richard L. Long, New Mexico State University, Las Cruces, NM

An interfacial area concentration balance has been derived from an interfacial enregy balance. To make computation tractable, simplfication of the source term is required. One approach is to use a modified order argument simplification. Application can be broadened to particulate flows based on an elastic cluster collision model.

See more of this Session: Novel Computational Methods in Mixing Research and Design
See more of this Group/Topical: North American Mixing Forum

Using Powder Surface Energies to Predict Mixing Properties of Model Excipient-API Blends

Daniel J. Burnett, Surface Measurement Systems, Allentown, PA

Ron Gray, Surface Measurement Systems, Allentown, PA

Nathaniel Catron, Abbott Laboratories, Abbot Park, IL

PURPOSE:

During pharmaceutical processing it is common to mix dry blends of different components. During transport or dispensing of this blend, the components may segregate. This can be particularly troublesome when the API is not dispensing uniformly. This can lead to dosage irregularities during production. Therefore, predicting blend stability of different mixtures can be valuable in optimizing product processing and formulation.

METHODS:

Binary blends of acetaminophen with the excipients were tested for initial goodness-of-blend (% RSD and drug loading) as well as blend stability after applied mechanical energy. Acetaminophen and excipients were sieved such that the largest acetaminophen particles were smaller than the smallest excipient particles. The blends were placed above a screen sized such that acetaminophen could pass through and tapped in order to promote drug loss if blends were not cohesive. Inverse Gas Chromatography (IGC) was used to measure the surface energy values of the Acetaminophen and excipients, independently. Then, the spreading coefficient was used to predict blend behavior.

RESULTS AND DISCUSSION:

The acetaminophen-mannitol formed a good initial blend and showed no loss upon tapping. Acetaminophen-microcrystalline cellulose formed a good initial blend, but showed significant loss of drug upon tapping. Acetaminophen-Prosolv showed a poor initial blend, but no significant loss upon tapping. The spreading coefficients (Acetaminophen over excipient) correlated well with the mechanical data. The spreading coefficients were as follows: 23.7, 18.2, and 8.6 mJ/m^2 for the Acetaminophen/mannitol, Acetaminophen/microcrystalline cellulose, and Acetaminophen/Prosolv mixtures, respectively. Higher spreading coefficients indicate the Acetaminophen is more likely to stick to the larger excipient particles.

CONCLUSIONS:

Both mechanical (% RSD and blend stability) and thermodynamic (spreading coefficient) mixing parameters indicated the following trend of excipients when blended with Acetaminophen: mannitol > microcrystalline cellulose > Prosolv. Therefore spreading coefficients obtained through the individual components' surface energy values could be used to predict ultimate blend performance.

See more of this Session: Research in Powder and Granular Mixing
See more of this Group/Topical: North American Mixing Forum

The Effect of Mixer Properties and Fill Level On Granular Flow in a Bladed Mixer

Brenda Remy, Dept. of Chemical and Biochemical Engineering, Rutgers University, Piscataway, NJ

Johannes G. Khinast, Institute for Process and Particle Engineering, Graz University of Technology, Graz, Austria

Benjamin Glasser, Dept. of Chemical and Biochemical Engineering, Rutgers University, Piscataway, NJ

Many industrial processes involving granular materials employ the use of mixers with mechanical agitators that provide shear, induce flow and encourage mixing. The appeal of these mixers comes from their ability to handle a wide variety of solids ranging from free flowing to cohesive powders and even pastes. Mixer characteristics and operating conditions determine degree of mixing and the shear profile achieved in such devices. Although these parameters play an important role, currently it is difficult - if not impossible - to predict granular behavior in mechanically agitated beds when these attributes are changed.

The discrete element method was used to study the effect of mixer properties and fill level on granular flow in a bladed mixer. Simulation velocity profiles were found to be in good agreement with experimental profiles obtained by using the Particle Image Velocimetry technique. For fill levels just covering the span of the blades, a 3-dimensional recirculation zone develops in front of the blades which promotes vertical and radial mixing. Increasing fill level reduces the size of the recirculation zone, decreases bed dilation and hinders particle diffusivities. However, above a critical fill level, the behavior of the particles within the span of the blade is found to be invariant of fill level. Cylinder wall friction is shown to significantly influence granular behavior in bladed mixers. Blade position along the vertical axis is shown to influence flow patterns, granular temperature and stress. The effect of increasing the mixer diameter at a constant particle diameter was also studied. When the mixer diameter is larger than a critical size such that wall effects are minimized, the observed granular behavior follows simple scaling relations. Particle velocities and diffusivities scale linearly with mixer size and blade speed.

See more of this Session: Research in Powder and Granular Mixing
See more of this Group/Topical: North American Mixing Forum

Novel Technique for Improving Granular Mixing in Rotating Cylinders

Suman K. Hajra, Computational Science Division, U. S. Department of Energy, National Energy Technology Laboratory, Pittsburgh, PA

D. V. Khakhar, Department of Chemical Engineering, Indian Institute of Technology, Bombay, Mumbai - 400076, India

Granular materials are widely used in industries such as cement, fertilizers, pharmaceuticals, construction, mining, and agricultural. Rotating cylinders are commonly used in the chemical and process industries as kilns, mixers, dryers, reactors, and granulators to process granular materials. Good cross-sectional mixing in the cylinders is important for product quality and efficient operation in terms of energy usage for continuous flow systems. Granular mixtures of particles with different properties (e.g., size, density) tend to spontaneously segregate. In this work, we study the radial segregation, transverse and azimuthal mixing in a rotating cylinder with an impeller. The radial profiles are obtained by sampling. Results are analyzed qualitatively with the profiles and quantitatively with mixing index, quantified in terms of segregation index which is a measure of relative proportion of small particles in the segregated core in an arbitrary segregated state to complete segregated state. Digital images are taken to qualitatively study the improvement of mixing for a cylinder with and without a rotating impeller. At a given impeller and cylinder speed, the extent of mixing was found to increase with increasing size of the impeller. For a given impeller size, the extent of mixing was found to increase with increasing speed of impeller rotation. No improvement of extent of mixing for regular pentagon as compared to circular geometry was observed.

See more of this Session: Research in Powder and Granular Mixing
See more of this Group/Topical: North American Mixing Forum

Mixing Patterns Obtained in An Oscillatory Shear Mixer

Luis Guillermo Obregon Quiñones, Chemical Engineering, University of Puerto Rico, Mayagüez Campus, Mayagüez, PR

Carlos Velazquez, Chemical Engineering, ERC for Structured Organic Particulate Systems, University of Puerto Rico, Mayaguez, PR

Even though there is a great amount of study of granular material, there still exists a great deal of interest in understanding concentrated granular flow because of its strong effect in the achievement of homogeneous granular mixing. Considering this emblematic problem in granular behavior, this work focused on developing a simple system where the imposed rate of strain was varied following a periodic function which permits to study the granular material random movement and the mixing of granular material at different wall amplitudes. The goal was to gain insights of the physical phenomena involved in the 3-dimensional granular flow using real powders with polydispersity on size. Results expose the presence of chaotic mixing in the superficial layer and are strongly related to the wall amplitude and particle size. The effect of the chaotic pattern is increased when increasing the velocity of the moving walls.

See more of this Session: Research in Powder and Granular Mixing
See more of this Group/Topical: North American Mixing Forum

Effect of Particle Size Distribution and Weight of Powders in Mixing of Granular Materials Under Periodic Shear

Luis Obregon, and Carlos Velazquez

Pharmaceutical Engineering Research Laboratory, Department of Chemical Engineering, University of Puerto Rico, PO Box 9046, Mayagüez, Puerto Rico 00681

Abstract

In this decade, the study of flow and mixing of granular materials had an exponential increase compared to the last decades. The advance has been noticeable in both computational and experimental studies. However, because of the large amount of factors that govern mixing and flow behavior, they continue being moderately understood. In this work, a new low shear mixing device consisting of a box with two moving walls and three static walls was developed to study granular mixing. The goal was to obtain a more in depth understanding of the internal behavior of granular material in the three dimensions when particle size distribution, imposed rate of strain, and initial filling height of the powders are changed. Results expose that the faster mixing is found when using particles of high diameter. The trend of mixing in the superficial layer differ extremely to the found in the bottom of the mixing chamber where a triangular pattern was obtained and whose size depended on the imposed strain rate, weight of the powders and particle diameter. The phenomenon of dilation was remarked varying the expansion of the bed linearly with the particle size

Keywords: Mixing; Granular Material; Shear; Particle Size

1 Introduction

The study of granular material behavior still is of major relevance for industries such as: chemical, metallurgical, pharmaceuticals, food, ceramics, etc.[1], and in geophysics in the study of turbidity currents, landslides, dry snow avalanches, pyroplastic flows, etc.[2].

Many different studies of blenders with interest for the pharmaceutical industry have been performed where different operating conditions such as shear stress, dimensions of the blenders, rpm in rotary blenders, have been varied. However, a complete understanding of the rheology of granular materials still is of concern in these applications where other factors such as particle density, wall-particle and particle-particle friction coefficient, restitution coefficient, angle of repose, angle of internal friction, young modulus, poison ratio, hardness, friability, internal porosity, cohesivity and particle size are present.

One of the most important factors in the study of granular flow and mixing is probably the effect of particle size and filling height. The majority of the work done about this issue has focused in segregation [3-5] and a few works have focused on the effect of changing the entire particle size in a system to study the distribution of particle. The latter works were performed in two dimensions and with monosized particles [6-9].

This paper expands the use of the oscillatory shear device to study the effect of particle size and filling height on the granular material flow. The work emphasized on applying a shear rate to granular material with specific particle size distribution and determining the random movement and the mixing of the granular material as a function of time. The goal was to gain insights of the physical phenomena (particle distribution, weight of powders, dilation) involved in the 3-dimension granular flow.

2 Experimental work

The experiments consisted in mixing two non-hygroscopic materials with similar properties, except the color, at different operating conditions in the mixing device that consisted of an open box with two moving walls and three static ones, as shown in Fig. 1. The width (W) or space between the moving walls was fixed at 2.79 cm. The length (L) and height (H) of the chamber were 22.86 and 11.43 cm respectively. The length of the moving walls (71.12 cm) was large enough compared to the length of the chamber (22.86 cm) to run experiments at different amplitudes of the periodic function.

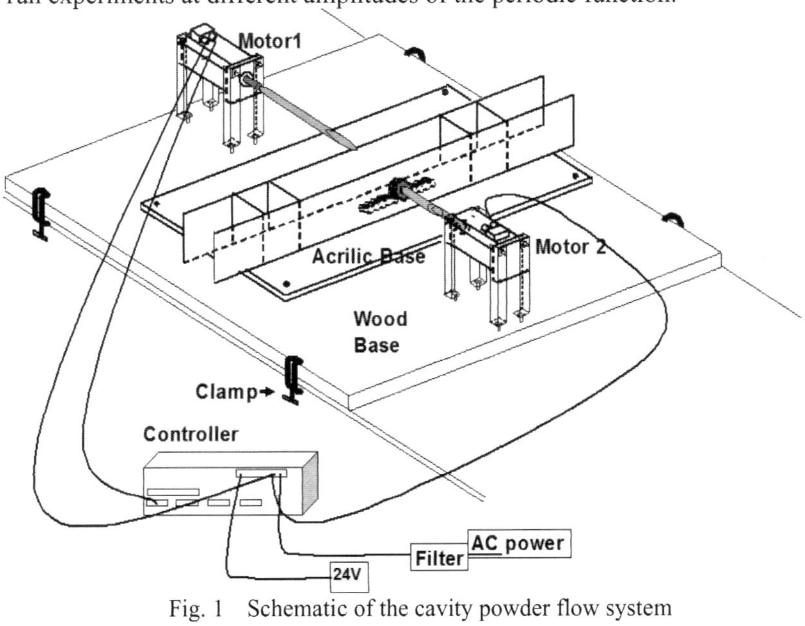

Fig. 1 Schematic of the cavity powder flow system

The wall motions were controlled by two step motors connected to a computer in a master-slave configuration. In one side of the chamber parallel to the moving walls it was poured granulated anhydrous lactose modified with a dark ink, and in the other side it was poured the normal granulated anhydrous lactose. They mixed under different rates of strain and the distribution of the particles tracked as a function of the time and space using a digital camera with high resolution and fast shutter speed. The camera was installed at the top of the equipment perpendicular to the flow field. As motion was applied to the walls the camera got periodic shots at the top surface. To study the movement in the three dimensions, a solidifier was poured slowly over the mixed powder at the end of the experiment. Then, the entire slab was cut vertically in 8 slices each with the same length Fig. 2. Pictures of each slice were taken to study the distribution of particles in the vertical layers (y and z axes).

Two operating conditions were considered in the study of the effect of particle size, amplitude of the periodic function and wall velocity. For the amplitude, it was used three levels, (11.43 cm, 22.86 cm, and 34.29 cm) and for the velocity it was used two levels, (13 cm/s and 16 cm/s).

To study the effect of Particle Size Distribution (PSD), The filling height of the granular material in the cavity was maintained constant in 5.08 cm.

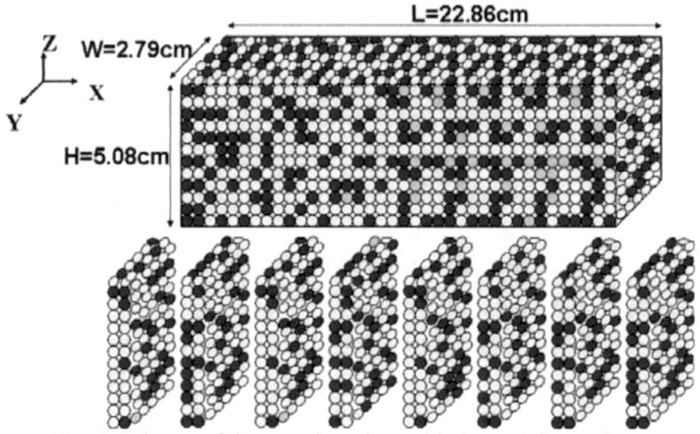

Fig. 2. Scheme of the samples taken with the real dimensions

The PSDs used were in the range of (600-425μm), (710-600μm), (850-710μm) and (1400-1000μm). Table 1 and Table 2 include the statistics of the particle size of both powders (dark and white). The distribution of particles was narrow enough as to avoid the phenomenon of segregation by particle size.

Two operating conditions were considered in the study of filling height, amplitude of the periodic function and wall velocity, each one with only one level, 34.29 cm and 13 cm/s respectively. The PSD used was in the range of (600 – 710 μm).

Table 1. Particle size distribution for blue particles in each range of study

Descriptive Statistics	**(600-425μm)**	**(710-600μm)**	**(850-710μm)**	**(1400-1000μm)**
Mean, μm	566.5	664.45	772.84	1251.68
Median, μm	550.5	663.98	778.00	1255.34
Standard Deviation	94.5	67.23	70.37	153.09
Kurtosis	1.3	0.60	0.44	-0.01
Skewness	0.5	-0.20	-0.10	0.06
Confidence Level	3.3	4.97	5.72	12.76

Table 2. Particle size distribution for white particles in each range of study

Descriptive Statistics	(600-425μm)	(710-600μm)	(850-710μm)	(1400-1000μm)
Mean, μm	571.7	655.73	735.63	1240.26
Median, μm	573.4	657.44	740.69	1247.14
Standard Deviation	103.7	87.78	74.20	163.47
Kurtosis	1	0.45	0.26	-0.22
Skewness	-0.3	-0.63	-0.24	-0.20
Confidence Level	4.4	8.62	5.83	16.15

The particle shapes were neither acicular nor flat, however, there is a lot of roughness in the surface. The powders exhibited no cohesiveness and thus flowed easily.

The function of the walls was as follow:

$$v = \begin{cases} a, cm/s \rightarrow if \ 0 \ \langle \ t, \sec \ \langle \ T/2 \\ -a, cm/s \rightarrow if \ T/2 \ \langle \ t, \sec \ \langle \ T \end{cases} \quad (1)$$

where T is the fundamental period.

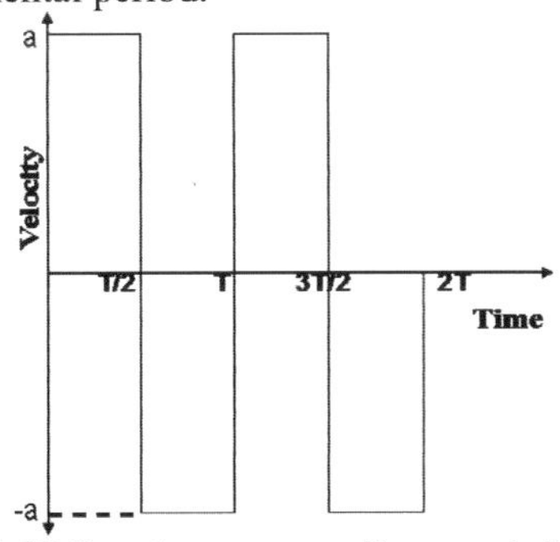

Fig. 3 Wall motions corresponding to periodic flow

3 Results and Discussion

3.1 Effect of the particle size on mixing of unrestricted powder at the top surface of the cavity

Fig. 4 is a top view of the cavity that shows how the particles were distributed in the surface (x, y) as motion was applied to the moving walls and as a function of time. The particle concentration in the surface was determined with a rectangular particle-counting technique. This technique divides the whole surface into n squares of the same size. Particle concentration is then measured by counting the number of white and blue particles inside each square domain [10, 11]. The letter P in Fig. 4 means the number of periods or cycles, where a period is the time the moving walls spent between leaving from a point and returning to the same point.

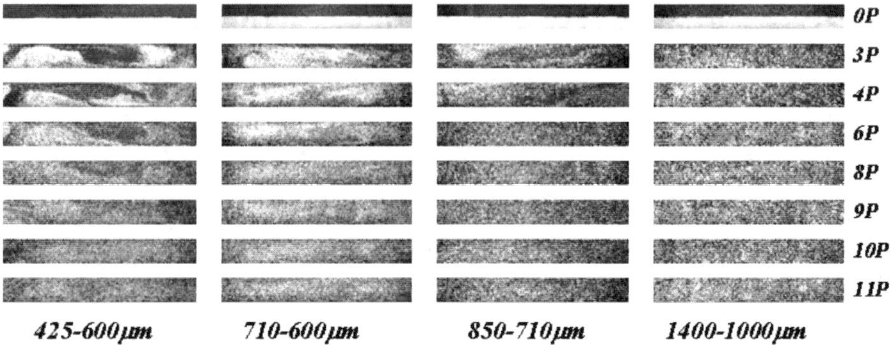

Fig. 4. Variation of mixing in the superficial layer (x,y axis). Effect of time and particle size. *v/W=4.65/s, A=34.29cm*

Since these surfaces were open to the air, the particles experienced zero friction and normal stress from above them allowing the particles to move more freely causing a higher transfer of momentum in this layer. In the case of layers with large diameter particles, the total friction under the particles is lower when compared with small diameter particles [8] due to the total surface area of particles.

The particles in the surface in contact with the moving walls had the same velocity regardless of the particle size. This behavior was attained due to the high roughness of the mentioned walls, which was constant for all the particle sizes. In addition, particles in the surface did not experience the effect of the weight of other particles, and considering that the particles of all the sizes in this study had similar physical properties including their shape because they were granulated in the same process with the same operating conditions, the variation in velocity due to particle properties should have been minimum. Hence the transfer of momentum in the surface from one line of particles to the contiguous line of particles should have happened in the same way for both, large and small particles. Therefore the reach of movement in the *y* axis away from the moving wall will depend on the particle size. The higher the particle sizes, the larger the y-distance of movement. For a given depth (y direction) smaller particles moved with lower velocity decreasing the transfer of momentum from its proportional y-distance of movement to the center of the cavity reducing as well the mixing in the surface.

Mixing curves of the surfaces were then generated by plotting the relative standard deviation (*RSD*) of the concentration as a function of mixing time (Number of cycles), Fig. 5, for each experiment. The *RSD* was computed as follow [12, 13]:

$$RSD = \frac{S}{\overline{C}} \quad (1)$$

where

$$S = \sqrt{\frac{\sum_{i=1}^{n}(C_i - \overline{C})^2}{n}} \quad (2)$$

S_i is the standard deviation, n is the total number of samples (squares) in the surface, $\overline{C}$ is the average concentration in the entire layer, and C_i is the concentration in the *i* sector.

It can be seen from Fig. 5 how fast the *RSD* curves reach their asymptotic value called the *RSD* for a random mixture. This is the best mixed state that the mixture can reach randomly and it is calculated as follow[14] :

$$RSD_r = \frac{S_r}{\overline{C}} \qquad (3)$$

where

$$S_r = \sqrt{\frac{C_w(1-C_w)}{m}} \qquad (4)$$

Cw is the overall proportion of white particles in the surface when the random mixing is reached and *m* is the number of particles in a sample.

The theoretical *RSDr* value (0.20) for these experiments is reached faster for higher particles and it is not exactly the same for all the sizes because of experimental differences. In addition, the treated particles had small differences in their sizes that could have caused segregation in a low scale producing a higher proportion of one of the powers in use in the surface. This small difference in proportion can only be detected by vision systems by means of the intensity of the pixels of the picture. Large scale segregation by density did not happen in these experiments because the difference in density between the two powders was lower than 0.3 g/cm^3 [10].

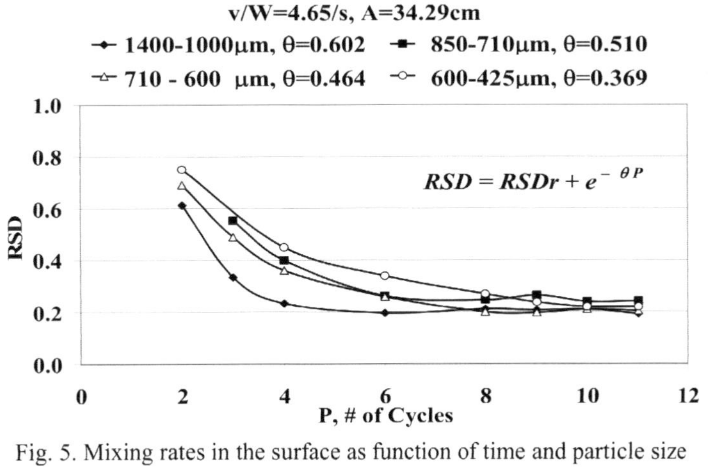

Fig. 5. Mixing rates in the surface as function of time and particle size

3.2 Effect of the particle size in the z-axis (height) of the cavity

Fig. 6 demonstrates the effect of the particle size on the mixing profile produced in the height direction of the cavity (plane *y-z*) at the center (*x=L/2*) of the cavity. It shows the distribution of particles moving from the walls (*y=0* and *y=W*) toward the center

($y=W/2$) as shear was applied. The triangle at the middle of the cavity is an area of no shear effect and thus no mixing. These triangles are important because they were formed due to the effect of the static wall at bottom where high friction dissipated the shear transferred from the moving walls and thus low mixing was obtained.

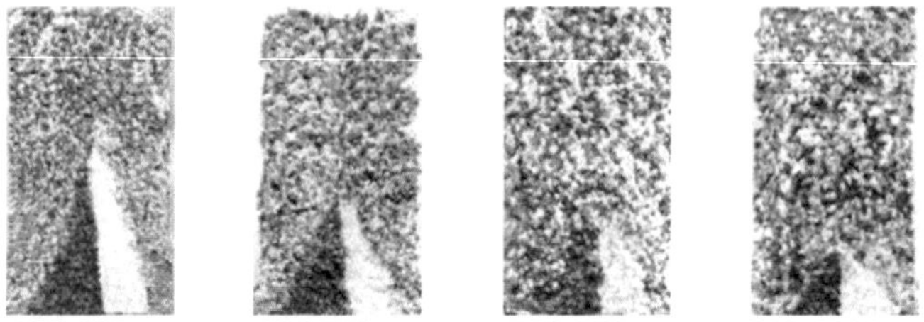

Fig. 6. Distribution of particles in the vertical layer at different particles sizes. ***v/W=4.65/s, A=34.29cm, t=60s***

In Fig. 6, it can be seen that there are two clearly demarked zones separated by a divisor line called slip line. These zones are the mixing zone and the static zone. The static zone (below the slip line) is where the particles did not move, thus they did not mix. The mixing zone, (above the slip line) is the zone of motion, which increased with increases in particle size, lowering the slip line. This expansion in the mixing zone occurred because of a better transfer of momentum from the moving walls to the center and bottom of the cavity. As it is stated in the figure above, the floor of the cavity remained static (highest friction because of the zero velocity and the effect of the weight of the particles), and the top remained opened to the air (lowest friction). This gradient in friction, together with the dilatancy (higher near the walls and lower in the center) produced a velocity profile in the three axes (x, y and z) that generated the particle distribution shown in Fig. 6. Particles in the compacted zone (center of the cavity) tended to move by diffusion to the dilated zone (toward the top and moving walls).

At the slip line, motion is extremely low because of the high friction offered by the static zone. This friction can be surpassed increasing the particle size as explained below. The density of all the particles (ρ_p) used was the same for every size, see Table 3, which makes that the weight of a vertical arrangement of particles with a specific height is higher for big particles than for small particles, even when there are fewer quantities of big particles in the mentioned arrangement. This effect of higher weight of large particles combined with the fact that the particles of all the sizes in this study had similar physical properties including their shape, caused an increase in the transfer of momentum in the z axis toward the bottom in the zone of higher porosity as a consequence of the dilation, increasing the diffusion in this direction as well.

Table 3, presents the relationship of the dilatancy (**ΔH**) with particle size for the white powder. The results for the blue powders were the same even when the difference in density between the two powders was 0.1 g/cm$^{3.}$ This dilatancy caused enough void fractions among particles and decreased the friction as well as to facilitate the vector movement of particles. It increased in turns the velocity to find the new equilibrium of forces among all the particles when shear forces were applied by the moving walls.

Fig. 7 presents specifically the dilation or expansion of the entire cavity proportionally to the particle diameter. The relationship is described by a slope of $\Delta H/D \cong 4.6$. This proportionality implies an increase in the effect of the particle weight in the motion of particles from the top to the bottom. The weight of a vertical array of particles increases with an increase in particle size. It indicates that for the moving wall to maintain the same velocity at the different particle sizes, it is necessary to increase the energy transmitted to the walls increasing as well the force applied to particles in the bottom of the cavity in order to maintain the increase in dilation which is proportional to the particle diameter. To maintain the dilation with the mentioned proportionality in diameter, the vertical component of the force applied by the roughness of the wall surfaces has to increase increasing in addition the horizontal component of the force causing a higher transfer of momentum in the y axis along the z axis producing a decrease in the triangle height improving thus the total mixing.

Table 3. Results of dilation of white powders in the cavity with a filling height of 5.08 cm and with a resulting volume of 294.97 cm^{3*}.

D (range) μm	*D* *mean*	**ΔH** *cm*	ρ_a *gr/cm³*	ρ_s *gr/cm³*	*Vs* *cm³*	*Vv* *cm³*	*Ve* *cm³*	*Vae* *cm³*	**ΔH/D**
1000 - 1400	1310.26	0.63	0.590	1.4	124.33	170.64	36.58	207.22	4.81
710 - 850	790.63	0.34	0.622	1.4	130.97	164.00	19.74	183.74	4.30
600 - 710	655.73	0.29	0.661	1.4	139.25	155.72	16.84	172.56	4.42
425 - 600	571.70	0.25	0.678	1.4	142.85	152.12	14.52	166.63	4.37

**D* : particle size, ΔH : expansion height, ρ_a : density of the bulk of powder in the cavity, ρ_s : density of the powders without void fraction, V_s : volume of the powders without void fraction, V_v : volume occupied by the void fraction at the initial time, V_e : expanded volume, and V_{ae} : total volume of the void fraction when powders are dilated

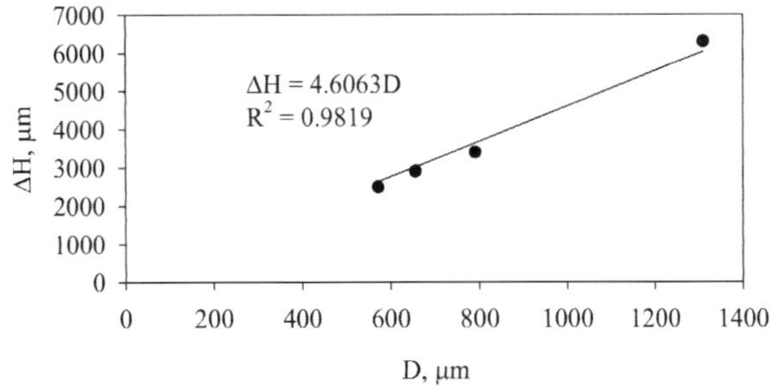

Fig. 7. Relation between the bed expansion of the powders in the cavity vs. the mean particle diameter. ***v/W=4.65/s, A=34.29cm, t=60s***

This complete study represents a nice illustration of how particle size distribution highly affects mixing practices for granular materials. However, much remains to be done with different powders to find more inclusive mixing characteristics.

3.3 Effect of the filling height in the z-axis (height) of the cavity

Fig. 8 shows four pictures of the mixing behavior at different initial filling heights. These pictures correspond to the depth of the cavity (plane *y-z*) at the center lengthwise, at $x=L/2$. This figure demonstrates the effect of the filling height on the produced mixing profile showing the distribution of the particles from the moving walls ($y=0$ and $y=W$) toward the center ($y=W/2$) as shear was applied. The higher the filling height the lower the triangle height producing a higher mixing degree. This suggests that the weight of the particles induces an increase in the transfer of momentum in the *z* axis.

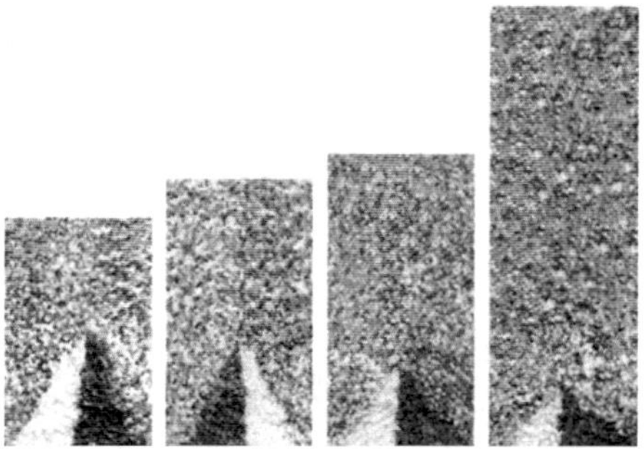

Fig. 8. Distribution of particles in the vertical layer at different filling heights, 3.81, 4.44, 5.08, and 7.62 cm from left to right. *v/W=4.65/s, t=60s, particle size (600-710 μm), Amplitude of wall=34.29 cm.*

The weight of a vertical array of particles increases evidently with an increase on the filling level. It indicates that for the moving wall to maintain the same velocity at the different initial filling levels, it is necessary to increase the torque of the shafts that move the walls Fig. 1. This increase in the torque helps to increase the energy transmitted to the walls increasing as well the force applied to particles in the bottom of the cavity in order to maintain the increase in dilation which is proportional to the height of the powder column.

4 Conclusions

It was confirmed that the small frictional area of large particles helps to distribute on the entire surface faster than small ones. The theoretical *RSDr* value (0.20) for these experiments is obtained faster for large particles and it is not exactly the same for all the sizes because of experimental error or segregation in a low scale producing a higher proportion of one of the powers in use in the surface.

Because the transfer of momentum in the surface from one line of particles to the contiguous line of particles should have happened in the same way for both, large and small particles, the reach of movement in the *y* axis away from the moving wall will depend on the particle size, so, the higher the particle sizes, the larger the y-distance of movement. In our case this area of movement is approximately 34 particle diameters. This area is different for different materials because depends on their physical properties.

In the depth of the cavity, the high friction occurring in the slip line that cause the no motion of particles, can be beaten increasing the particle size due to the increase of weight of the vertical columns of particles increasing the transfer of momentum in the z and y axes as a consequence of the dilation which was found to be proportional to the particle diameter in the entire cavity with the expression $\Delta H/D \cong 4.6$.

The size of the triangle cannot be used to determine the degree of homogeneity of the system but it can be used to determine the highest level of mixing the entire cavity can obtain, which is just when the mixing zone reach the random mixing.

The trend of the triangle height with the initial filling height reveals that the zone of movement in the surface not only depend on the particle size but also on the filling height.

References

1. Jain, N., J.M. Ottino, and R.M. Lueptow, *Effect of interstitial fluid on a granular flowing layer.* Journal of Fluid Mechanics, 2004. **508**: p. 23-44.
2. Dartevelle, S., *Numerical modeling of geophysical granular flows:1. A comprehensive approach to granular rheologies and geophysical multiphase flows.* Geochemistry Geophysics Geosystems, 2004. **Volume 5**(8): p. 1-28.
3. Cisar, S.E., P.B. Umbanhowar, and J.M. Ottino, *Radial granular segregation under chaotic flow in two-dimensional tumblers.* Physical Review E - Statistical, Nonlinear, and Soft Matter Physics, 2006. **74**(5): p. 051305.
4. Ottino, J.M. and D.V. Khakhar, *Mixing and segregation of granular materials.* Annual Review of Fluid Mechanics, 2000. **32**: p. 55-91.
5. Yang, S.C., *Density effect on mixing and segregation processes in a vibrated binary granular mixture.* Powder Technology, 2006. **164**(2): p. 65-74.
6. Campbell, C.S., *Boundary interactions for two-dimensional granular flows. Part 1. Flat boundaries, asymmetric stresses and couple stresses.* . Journal of Fluid Mechanics, 1993. **247**: p. 111-136
7. Hajra, S.K. and D.V. Khakhar, *Radial mixing of granular materials in a rotating cylinder: Experimental determination of particle self-diffusivity.* Physics of Fluids, 2005. **17**(1): p. 013101-013101.
8. Iordanoff, I., K. Elkholy, and M.M. Khonsari, *Effect of particle size dispersion on granular lubrication regimes.* Proceedings of the Institution of Mechanical Engineers, Part J: Journal of Engineering Tribology, 2008. **222**(6): p. 725-739.
9. Tardos, G.I., S. McNamara, and I. Talu, *Slow and intermediate flow of a frictional bulk powder in the Couette geometry.* Powder Technology, 2003. **131**(1): p. 23-39.
10. Obregon, L. and C. Velazquez, *Discrimination limit between mean gray values for the prediction of powder concentrations.* Powder Technology, 2007. **175**(1): p. 8-13.
11. Realpe, A. and C. Velazquez, *Image processing and analysis for determination of concentrations of powder mixtures.* Powder Technology, 2003. **134**(3): p. 193-200.
12. Obregon, L., et al., *Mixing of Granular Materials: Part I - Effect of Periodic Shear.* Powder Technology 2009. **Accepted for publication**.
13. Arratia, P.E., et al., *A study of the mixing and segregation mechanisms in the Bohle Tote blender via DEM simulations.* Powder Technology, 2006. **164**(1): p. 50-57.
14. Lacey, P.M.C., *The mixing of solid particles.* Chem. Age, Lond., 1945. **119**(145): p. 53-59.

Water Quality Analysis Related to Finished Water Mixing in Water Storage Tanks

Jason A. Heberling, Department of Civil and Environmental Engineering, University of Alabama at Birmingham, Birmingham, AL

The degradation of finished water in storage tanks is of great concern to water utilities. It is widely assumed that tank mixing enhances water quality, and slows the creation of disinfectant byproducts (DBP's). In an attempt to enhance water quality and to anticipate future, tighter regulations on DBPs, the Birmingham Water Works Board (BWWB) is conducting a two-part study to examine storage tank mixing effectiveness by comparing tanks with turbulent jet mixing devices and ones without any mixing enhancement.

Part 1.

Two tanks of identical volume, one equipped with a mixing device and the other sporting no such modification, are compared in this study. The two tanks, located less than sixty feet apart, are supplied finished water from the same source at essentially the same pressure. Both tanks maintain roughly identical volumes as the tanks fluctuate through the day and both tanks are fitted with temperature sensors spaced at equidistant heights inside the tanks. Fluoride will be used as a tracer to measure mixing rates. Temperature, fluoride tracer concentration, and disinfection byproducts concentration will all be measured from different elevations in the tank. In addition, the relationship between drawdown and mixing rates will be examined by sampling the tanks elevation ranges from peak demand (lowest volume) to full volume. The data collected will give a clear picture of the effectiveness of the mixing apparatus as compared to the tank with no such apparatus. The data is collected over a twelve (12) month period to account for seasonal variation.

Part 2

Additional tanks throughout the distribution system utilize mixing devices in an attempt to mix stored water and reduce water age. These devices are found in all tank types (standpipe, reservoir and elevated). Nine (9) storage tanks scattered throughout the BWWB distribution system, sporting mixing devices, will be sampled for temperature, free chlorine, total haloacetic acid (HAA's), and total trihalomethanes (TTHM's) at varying heights. These parameters will be measured over a twelve (12) month period to account for seasonal variation. A properly mixed tank should show little variation top to bottom of the sampled parameters.

The combined information from this project should give a detailed comparison between tanks with mixers and tanks without mixers thus demonstrating the effectiveness of the turbulent jet mixers. Data will be analyzed using Systat (Version 11), Sigmaplot (Version 10), and Microsoft Excel.

See more of this Session: Mixing in Water Purification and Waste Treatment / Applied Mass Transfer
See more of this Group/Topical: North American Mixing Forum

Impeller Design for Anoxic Mixing Services

Wojtek Wyczalkowski, R&D, Philadelphia Mixing Solutions LTD, Palmyra, PA

Bob Higbee, R&D, Philadelphia Mixing Solutions LTD, Palmyra, PA

Binxin Wu, R&D, Philadelphia Mixing Solutions LTD, Palmyra, PA

Mixing of anoxic basins in Waste Water Treatment Plants require both a relatively uniform suspension of solids and good blending without disturbing the liquid surface. In order to decompose nitrogen compounds, bacteria must be starved of dissolved oxygen which forces the bacteria to consume the chemically bound oxygen within the nutrients. Hydrofoil impellers are typically used for such service. Since energy costs have increased recently, and will increase further in the near future, impeller efficiency is a key issue. A large waste water plant may have between 30-80 anoxic mixers which will consume a significant amount of energy. This presentation discusses an impeller design that is fabricated using composite plastic materials. The choice of composite plastic materials allows one to have 100% control over the blade geometry, unlike typical industrial hydrofoil impellers, which are made by forming flat steel plate. The composite plastic impeller geometry utilizes novel circular rake technology for improved pumping efficiency. Blades are also hyper-skewed in order to avoid the collection of rags. The impeller performance is reviewed and compared to different impeller technologies used for anoxic mixing in terms of blending and solids suspension.

See more of this Session: Mixing in Water Purification and Waste Treatment / Applied Mass Transfer
See more of this Group/Topical: North American Mixing Forum

Water and Wastewater Mixing

Michael K. Dawson, BHR Group Limited, Cranfield, United Kingdom

David A. R. Brown, BHR Group, Cranfield, United Kingdom

This paper discusses a number of mixing processes employed in the production of potable water and the treatment of waste-water (including sewage sludge applications). Methods of obtaining and correlating appropriate performance data for design and scale-up purposes are presented.

See more of this Session: Mixing in Water Purification and Waste Treatment / Applied Mass Transfer
See more of this Group/Topical: North American Mixing Forum

Low-Order Modeling for the Effective Cleaning Radius of a Pulse Jet Mixer

L. J. Peltier, Advanced Simulation and Analysis, Bechtel National, Inc., Frederick, MD

YoChan Kim, Advanced Simulation and Analysis, Bechtel National, Inc., Frederick, MD

Henry Fitzpatrick, Advanced Simulation and Analysis, Bechtel National, Inc., Frederick, MD

Kelly Knight, Advanced Simulation and Analysis, Bechtel National, Inc., Frederick, MD

Brigette Rosendall, Advanced Simulation and Analysis, Bechtel National, Inc., Thousand Oaks, CA

Chris Kennedy, Advanced Simulation and Analysis, Bechtel National, Inc., San Francisco, CA

Jon Berkoe, Advanced Simulation and Analysis, Bechtel National, Inc., Frederick, MD

Pulse Jet Mixers are designed at the Hanford Site Waste Treatment Plant to keep radioactive solids in suspension prior to vitrification for permanent storage. These systems must operate dependably and without maintenance for 40 years or more. The existing design database for pulse jet mixing of slurries like those at Hanford is meager. Verified/validated low-order design tools do not exist that can be used to reliably predict the operating conditions under which the Hanford slurry will successfully achieve suspension.

This study presents the results of a coupled CFD/analytical effort to develop a low-order design tool to support the design/evaluation of pulse jet mixers. The modeling begins with the Eulerian-Granular 2-phase flow equations governing the suspension, transport, and deposition of particulates in a carrier fluid. The equations are simplified based on observations from CFD pertinent to the zone-of-influence for the pulsed jet flow leading to a reduced set for which the solution relating the wall shear-stress to the radial distance from the pulse jet impingement point is known. By evaluating this expression at the critical shear stress for solid suspension, a model for the effective cleaning radius is evolved.

The model is validated using comparisons to CFD results and by comparisons to empirical correlations derived from experimental data.

See more of this Session: Mixing in Water Purification and Waste Treatment / Applied Mass Transfer
See more of this Group/Topical: North American Mixing Forum

Novel Design for a Magnesium-Enhanced Lime Ex-Situ Gypsum Oxidizer

Chris Hibshman, Philadelphia Mixing Solutions LTD, Palmyra, PA

Bob Dowd, CA, Philadelphia Mixing Solutions LTD, Palmyra, PA

Wojtek Wyczalkowski, R&D, Philadelphia Mixing Solutions LTD, Palmyra, PA

Kevin Smith, Carmeuse Lime and Stone, Pittsburgh, PA

Carl Larid, R&D, Carmeuse Lime and Stone, Pittsburgh, PA

Julia Darling, R&D, Carmeuse Lime and Stone, Pittsburgh, PA

One method of gypsum production in FGD systems is magnesium-enhanced lime (MEL) ex-situ oxidation. A typical 1400MW commercial system consists of two 9.1 m (360") diameter reactors with 13.0 m (516") liquid level. An air sparge grid is the sole means of agitation and solids suspension. This existing design is proven to be effective but is not optimized for agitation. For example, more air may be added to achieve agitation than is required for oxidation. This paper presents the novel design of a mechanically agitated ex-situ gypsum oxidizer/crystallizer. Presented data was collected on 0.6 m (24") diameter pilot installation of countercurrent oxidizer/crystallizer consisting of four separate mixing stages. Calcium sulfite from a commercial MEL scrubber was oxidized to form gypsum crystals. Pilot scale results demonstrate a significantly reduced stoichiometric oxygen requirement. Calculations for a commercial scale system show a 50% reduction of compressor power. Combined with the power requirement for agitation the net energy savings may be in the range of 20-30% over the existing installation.

See more of this Session: Mixing in Water Purification and Waste Treatment / Applied Mass Transfer
See more of this Group/Topical: North American Mixing Forum

Improved Mass Transfer Efficiency in Wet Limestone Flue Gas Desulfurization

Chris Hibshman, Philadelphia Mixing Solutions LTD, Palmyra, PA

Bob Dowd, CA, Philadelphia Mixing Solutions LTD, Palmyra, PA

Wojtek Wyczalkowski, R&D, Philadelphia Mixing Solutions LTD, Palmyra, PA

The most commonly used Flue Gas Desulphurization systems use limestone forced oxidation in order to produce fully oxidized gypsum products. Oxidation air is dispersed into the solution, typically by means of injection in front of a rotating impeller.

Efficiency of the oxidation is determined by the value of the mass transfer coefficient (kLa) and the system resistance to impeller flooding. High mass transfer coefficient may help to reduce the size of the air compressors. The system resistance to flooding allows for close proximity of the injection point to the impeller and further increases the mass transfer coefficient.

The paper discusses the new design of the air injection technology that increases mass transfer coefficient up to 30%, compared with a traditional lance injection. In addition, the new system is very resistant to flooding, and it is perfectly suitable for cases where high air flow is required.

See more of this Session: Mixing in Water Purification and Waste Treatment / Applied Mass Transfer
See more of this Group/Topical: North American Mixing Forum

AUTHOR INDEX

Akiti, Otute 89
Albaek, Mads O. 91
Allen, Billy R. 85
Allotta, Terrence 92
Alvarez, Mario M. 93, 138
Amanullah, Ashraf 92
Armenante, Piero 1, 14, 85, 88, 149
Berkoe, Jon 173
Berson, R. Eric 145
Brown, David A. R. 13, 29, 172
Bujalski, Waldek 16
Bulnes-Abundis, David 93, 138
Burnett, Daniel J. 156
Calabrese, Richard V. 17, 151
Caramellino, Micaela 14, 88
Carrillo-Cocom, Leydi Maribel 138
Castillo-Reyna, Josefina 93
Catron, Nathaniel 156
Chomcharn, Nonjaros 85
Chu, Liang-Yin 33
Cooke, Michael 16
Cusack, Sean P. 89
Darling, Julia 174
Dawson, Michael K. 13, 172
Delmotte, Yves 103
Derfus, Gayle E. 92
Derksen, Jos 150
Di Benedetto, Giuseppe 149
Donley, Rhea 89
Dordevich, Milorad 148
Dowd, Bob 55, 174, 175
Duvvuri, Subbarao 18
Etchells III, Arthur W. 56
Fasano, Julian B. 34, 73
Fitzpatrick, Henry 173
Garcia-Barberena, Maria 56
Gezork, Klaus 16
Gilchrist, James F. 68
Gillissen, Jurriaan J. J. 153
Glasser, Benjamin 157
Granados, Marissa 93
Gray, Ron 156
Hajra, Suman K. 158
Hammad, Khaled J. 2
Hanley, Thomas R. 86
Hasebe, Shinji 152
Heberling, Jason A. 170
Hesketh, Robert 56
Hewitt, Christopher J. 90
Hibshman, Chris 55, 174, 175
Higbee, Bob 12, 171
Hsu, Ruey Chi 72
Hudson, Terry 92
Iftikhar, Muhammad Usman 56
Janz, Eric E. 34, 73
Jiang, Yi 70
Joaquim Jr., Celso F. 15
Johnson, Keith 138
Kano, Manabu 152
Kar, Kishore 69
Kennedy, Chris 148, 173
Kennedy, David J. 86
Khakhar, D. V. 158
Khinast, Johannes G. 157
Khopkar, Avinash 69
Kiger, Kenneth T. 17
Kim, Yochan 148, 173
Knight, Kelly 173
Koopman, David C. 48
Kothary, Kapil 92
Kresta, Suzanne 15
Lambert, Daniel P. 48
Larid, Carl 174
Lester, Daniel 146
Long, Richard L. 155
Lwin, Ye 18
Macgregor, Cameron A 56
Machado, Marcio B. 15
Madenjian, Ed 69
McAdams, Todd A. 102
McLeod, Georgina 90
Metcalfe, Guy 146
Milanovic, Ivana M. 2
Morris, Jeffrey F. 58
Moseley, Marc 147
Motamedvaziri, Shilan 1
Murthy, B. N. 17
Murugappan, Gayathree 138
Myers, Kevin J. 34, 73
Myint, Thomas 92
Nienow, Alvin W. 16, 90, 91
Nobes, David 15
Nunhez, Jose Roberto 15
Ooms, Gijs 30
Ozcan-Taskin, Gul Nerime 57
Padron, Gustavo A. 57
Parmar, Nilesh 69
Patterson, Gary K. 154
Peltier, L. J. 173
Pickenheim, Bradley R. 48
Piela, Krzysztof 30
Quiñones, Luis Guillermo Obregon 159, 160
Reid, Charles 59
Remy, Brenda 157
Rosendall, Brigette 148, 173
Rudman, Murray 146
Sable, Dhanshri 18
Sajja, V. S. K. 86
Sengers, Jan V. 30
Sharp, M. Keith 145

AUTHOR INDEX

Smith, Kevin 174
Smith, Philip J. 59
Sokolnicki, Adam M. 87
Somasi, Madan 69
Stocks, Stuart 91
Stone, Michael E. 48
Thomas, Jonathan 145
Thornock, Jeremy N. 59
Todd, Paul W. 86
Tolentino, Deana 89
Tonomura, Osamu 152
Van Den Akker, Harry E. A. 153
Velazquez, Carlos 159, 160
Wang, Lin 152
Weigand, Keith 56
Wu, Binxin 12, 171
Wu, Chung Tung 72
Wyczalkowski, Wojtek 12, 55, 147, 171, 174, 175
Xu, Bu 68
Yang, Chao 70
Yang, Meng 151
Yong, Yumei 70
Zheng, Han 149